Mechanical Deburring and Surface Finishing Technology

MANUFACTURING ENGINEERING AND MATERIALS PROCESSING

A Series of Reference Books and Textbooks

SERIES EDITORS

Geoffrey Boothroyd

*Chairman, Department of Industrial
and Manufacturing Engineering
University of Rhode Island
Kingston, Rhode Island*

George E. Dieter

*Dean, College of Engineering
University of Maryland
College Park, Maryland*

OTHER VOLUMES IN PREPARATION

Mechanical Deburring and Surface Finishing Technology

ALFRED F. SCHEIDER

Osborn Manufacturing/Jason Incorporated
Cleveland, Ohio

Marcel Dekker, Inc.   **New York and Basel**

Library of Congress Cataloging-in-Publication Data

Scheider, Alfred F.
    Mechanical deburring and surface finishing technology / Alfred F.
Scheider.
        p.   cm. – (Manufacturing engineering and materials processing
; 35)
    Includes bibliographical references and index.
    ISBN 0-8247-8157-0
    1. Deburring. 2. Surfaces (Technology) 3. Finishes and
finishing. I. Title. II. Series.
TJ1280.S34   1990
671.7–dc20   90-36790
                                                            CIP

This book is printed on acid-free paper.

MARCEL DEKKER, INC.
270 Madison Avenue, New York, New York 10016

Current printing (last digit):
10 9 8 7 6 5 4 3 2 1

PRINTED IN THE UNITED STATES OF AMERICA

Preface

Significant changes have taken place in recent years in mechanical deburring, edge contouring and surface finishing tools, methods, processes, and systems. With the rapid evolution of new materials used in the manufacture of component parts, near net shape parts, and advanced composites, conditions of increased surface toughness, hardness, and high abrasion developed that conventional brushing and buffing tools could not handle.

In addition, the dimensions and tolerances of component parts became increasingly critical, with shapes and geometry exhibiting complexity never experienced before. At the same time, robotics, automated machinery, and machine tools with their sophisticated computer numerical control systems were bringing a new dimension to the materials removal industry with advancements in productivity, quality, and cost effectiveness. It made sense for the manufacturer to complete all the machining operations and the final finishing—deburring, edge contouring, and surface conditioning—during the programmed cycle.

The combination of the machining and finishing operations eliminated the need for a secondary manual or semiautomatic finishing of the component parts.

As recently as three years ago few brush and buff tools had been developed that could interface with advanced machine tools to finish parts made of new materials with critical dimensions, shapes, and geometries. This brings us to the importance of this book.

The writing of the original manuscript commenced two years ago, at a time when many new flexible abrasive finishing tools were being developed, along with the application of new technologies. To ensure that the reader is exposed to all facets of mechanical deburring and surface finishing technology, the book is divided into six chapters. The Introduction is a historical overview based on my more than 30 years in the industry, all with my present company, responsible for the engineering and scientific professional fields.

The middle chapters focus the reader's attention on both basic and sophisticated engineered materials, product and construction design technologies, and many unusual and unique shop floor product applications and supporting technical data.

The last chapter has the most important and informative material for engineers, technicians, and supervisory management personnel in manufacturing engineering, tool engineering, and manufacturing research and development as it highlights the new technologies that have been developed during the last few years. I also discuss future trends and how they will develop into new state-of-the-art finishing tools to be used in the global market during the 1990s.

Alfred F. Scheider

Contents

Mechanical Deburring and Surface Finishing Technology

1

Introduction

Mechanical deburring and surface finishing with power-driven brushes and buffs had its serious beginnings in the mid-1940s with the advent of bench and pedestal floor grinders, along with elementary portable hand-held air and electric power tools.

Prior to that time, brushes and buffs were manually held and applied to the work with human muscle and hand strength. A typical example of the early methods in the steel fabricating industry, where welded joints were made, was to break up the flux and scale with a hand-held chipping hammer and follow up manually with a wood-backed wire hand scratch brush to clean the loose and loosely adhering debris away.

Similar examples were applied in the buffing area, where buff materials with cutting and coloring compounds were used to achieve metal finishes by tedious and laborious hand rubbing and wiping.

MECHANICAL OPERATIONS

By the late 1940s and early 1950s, as industry geared its factories and manufacturing facilities to the high production of consumer goods, a great deal of emphasis and focus was placed upon mechanizing operations by adding semiautomatic machinery and equipment to replace the hand labor operations. This change in direction toward mechanization created a need for good quality mechanical power-driven brushing and buffing tools that did not exist, and these tools were developed on a rapid basis to meet the ever growing needs of industry.

As new turning, boring, milling, shaping, drilling, and other special machinery was developed and put into operation, a secondary operation was created to remove burrs (deburring) created by the primary machining and metal removal operations. Also, sharp edges needed to be blended, rounded, or radiused (edge contouring) to design-engineered specifications. Flat surfaces and contoured ones needed to be cleaned, smoothed, polished, buffed to a high luster, and, in some cases, roughed for providing additional surface area prior to the bonding of other materials (surface conditioning).

The combination of new state-of-the-art production machinery, equipment, tooling, and power brush and buff tools provided high-volume productivity and good, consistent quality parts, manufactured on a reliable and predictable basis, and easily assembled into component assemblies and/or final products.

From the early 1950s through the late 1960s, a tremendous amount of development work was done by many brush and buff companies in the area of fill (filler) materials, such as wire, synthetics, natural fibers, cloth, and abrasive compounds, as well as on design and construction of various types, sizes, shapes, and geometries of the finished products. This evolution took place to meet the industry demands for their ever-increasing and changing parts and components design improvements and to meet the product application requirements as they became more critical. Besides brush and buff companies, there were other competitive methods for companies, such as mass media finishing, coated and flexible abrasives, carbide tools, and other special processes being perfected, competing on the basis of consistent quality, lower end of service costs, and productivity. During this time frame, mechanical finishing evolved into an integral part of many primary manufacturing operations, casting aside the misnomer of being only a secondary operation that was performed when needed and as an afterthought often overlooked by the design engineering and manufacturing group. Illustrating the point is the automotive transmission gear manufacturing process, which, during this time, added wire brushing to deburr and edge-blend gear tooth profiles after the gear cutting operations. Both are primary operations and necessary for the consistent-quality, trouble-free, high-performance transmissions.

SOPHISTICATED MATERIALS AND PROCESSES

Enter the 1970s, with its rapidly growing electronics industry and applications, sophisticated new high-technology metals and materials, and improved manufacturing systems processes, which accelerated the changes in brush and buff technologies, applications, fill materials, and tool design and construction.

Emerging developments in the areas of plastics, comprising composite and advanced composites materials (favored particularly for the ratio of high strength to light weight), found their way into the aerospace, aerostructures, aircraft engine, and automotive industries. This created new problems and opportunities for the mechanical finishing industry. Entire new concepts, applications, and tool products were developed to replace the tools used on applications replaced by the composite materials.

An extremely important brush fill material developed during this period was the abrasive/nylon monofilaments that can be made into flexible abrading and grinding tools to conform to flat, interrupted, and contoured surfaces of metallic and nonmetallic parts. Buffs comprised of cotton and synthetic fabrics with abrasive grains attached with adhesive were also under development during this time.

RECENT DEVELOPMENTS

Since the early 1980s there has been a proliferation of advanced, sophisticated, precision, computer-controlled machinery, equipment, systems, and processes, focusing not only on the four major material removal methods of turning, drilling, milling, and grinding, but also all others.

Separate deburring, edge contouring, and surface finishing machinery, equipment, robots, cells, systems, and processes, complete with computerized controls and handling systems, enabled industry to minimize much of the hand deburring still being done in the deburring department or on ''burr benches'' in some shops.

Most recently, flexible abrasive finishing tools, designed for deburring, edge contouring, surface conditioning, and cleaning, have been integrated into NC (numerical controlled) and CNC (computer numerical controlled) machine cycles, directly following the turning, drilling, milling, and grinding operations. With the multiplicity of tool holders in the magazine, cutting and finishing tools can be accommodated to completely machine, finish, and inspect the piece parts in one machine cycle operation without any off-machine operations or people needed.

The success of these new recent developments is attributed for the most part to the design, development, and engineering of flexible abrasive finishing tools, with very little activity in conventional wire and natural fiber brushes, buffs, and abrasive compounds. The rationale is based on the use of abrasive and superabrasive minerals, which can remove amounts of any material. The tools are made with greater accuracy and concentricity, and with exact balance, uniform predictability, and positive reliability in performance.

With the dynamic technological changes continually taking place in industry worldwide, what worked well five years ago in the environment of that time is not acceptable in today's state-of-the-art conditions.

New products, tools, materials, ideas, and creativity are being generated by most tool manufacturers today and will continue through the rest of this century and beyond. It is absolutely essential to the survival of our manufacturing industries, because of the competitive nature of the global marketplace.

2

Cutting and Grinding Fill Materials

Industrial power-driven brushes and buffs used for mechanical deburring and surface finishing are inherently flexible tools that depend on the fill materials, which do the work, to bend and recover within their elastic limits. Contacting flat or irregular contoured surfaces, the rotating fill materials continually strike, impact, wipe, cut, and grind the work—then instantaneously recover—only to repeat the process cycle again and again until the tool reaches the end of its life cycle and is replaced.

Selection of the proper fill materials is the single most important criterion in achieving the performance results desired at the lowest end-of-service costs expected.

Although some of these fill materials are man-made and others are from natural sources, they are known collectively as *brush or buff fill material*. These materials include tempered and untempered high-tensile steel wires, annealed stainless steel wires, and a number of nonferrous wires such as brass, nickel silver, beryllium copper, titanium, and zirconium. Other steel wires are brass, zinc, and/or plastic coated, basically for appearance or corrosion resistance.

Abrasive monofilaments, clustered abrasive monofilaments, plastic-coated fiberglass, nylon, polypropylene, polyester, and carbon/graphite lead the growing list of man-made fill materials.

All of the man-made fill materials for power brushes are provided in two basic forms. The first form is *straight*; the other form is *crimped*. Crimped fill material is produced by processing the straight material through two gear sets (offset by 90 degrees), thereby introducing a predetermined, consistent, and uniform amplitude and frequency. Figure 2.1 illustrates this.

Natural fiber, which is to say, materials grown by nature, include tampico, sisal, bass, bassine, bahia, palmetto, palmyra, cotton, linen, and others of less consequence. These make up the majority of fill materials. In some cases, specialized combinations of fibers are used for specific applications by mixing together portions of each basic fiber.

Carbon steel wire has been known and used in industry for over 50 years. It was not until the late 1940s and early 1950s that the development of high-tensile oil-tempered brush wire took place. Over the years, the much improved fatigue resistance and superior cutting characteristics of the wires accounted for the rapid growth of industrial power-driven brushes.

STEEL WIRE

Tempered Carbon Steel Wire

Tempered carbon steel wire is still used in power brushes more so than any other fill materials, with the tempered predominating over the untempered, hard-drawn variety. This steel wire is high-quality material similar to that used in springs and other special engineered products. It is generally made from 0.68% to 0.75% carbon steel rod with tensile strengths ranging from 320,000 to 380,000 psi (pounds per square inch) developed at the wire drawing mills. Because of the millions of cycles of compressive and linear stresses to which the filaments are subjected, it is

CRIMPED WIRE

STRAIGHT WIRE

Figure 2.1   Basic forms of man-made fill materials.

very important that the wires be able to withstand repeated flexing without premature fatigue and breakage. The wire must be bright finished and free of pits, die marks, rust, scale, scrapes, splits, laps, cracks, seams, and excessive decarburization. A grade of tempered wire with lower tensile strengths ranging from 180,000 to 320,000 psi and 0.60% to 0.70% carbon content is also used where the resistance to abrasion and the service requirements are not as demanding.

Untempered Carbon Steel Wires

Untempered carbon steel wires are hard drawn to their highest tensile strength and are usually coppered or liquor finished. The solution-drawn wire has an extremely smooth, thin, residual finish coating. Three grades are normally used.

1. Low-carbon brush wire (0.14% to 0.20% carbon) with a tensile strength of approximately 140,000 psi.

2. Hand-scratch brush wire (0.45% to 0.60% carbon) with a tensile strength of 230,000 to 290,000 psi, depending on the diameter of the wire.

3. High-strength wire (0.60% to 0.75% carbon) with a tensile

strength of 300,000 to 380,000 psi, depending on the diameter of the wire.

The low-carbon brush wire is commonly produced in sizes of 0.002 to 0.006 inches in diameter, while hand-scratch brush wire and high-strength wire are usually produced in diameters ranging from 0.006 to 0.035 inches in diameter. Untempered wires do not have a great cutting capability or consistency of physical attributes, nor are they as abrasion-resistant as tempered wires. However, they are lower in cost and are usually adequate for the purposes intended.

Stainless Steel Wire

Stainless steel wire is used primarily for corrosive environments, elevated temperature applications (over 350°F), and where the avoidance or minimization of carbon deposits would be helpful to the work parts being brushed (nontransfer of carbon deposits that cause oxidation).

The most suitable brush wire is bright-finished AISI type 302, which is drawn and annealed to minimum tensile strengths of 320,000 psi. The higher the tensile strength, the better the performance, from a deburring, cutting, or cleaning standpoint. AISI type 304 (lower carbon content) is also suitable as an alternate selection, providing that the tensile strength minimum of type 302 is met or exceeded.

On limited occasions, ANSI type 316 is used for brush fill materials because of its characteristics to better withstand highly corrosive and elevated temperature application environments.

In the austenitic grade of stainless steel wire, AISI type 420 can be used where the possible transfer of carbon is not a hindrance to the work being processed. Applications such as wet scrubbing with water or some chemicals would be a good workable example. Having higher carbon content than type 302, the type 420 steels can be tempered for higher tensile properties, thereby giving better life and cut performance. Yet with these advantages it does not come close to matching the life and cut performance of the high-tensile-strength carbon steel wire discussed earlier.

There are some applications where it is desirable to use brushes with nonmagnetic stainless steel wire. Untempered type 302 steel, when drawn to quality used in brushes, can be attracted by a magnet to varying degrees. The amount of magnetic attraction for a given size of wire in the tensile range is a function of the degree of cold working that the wire has undergone. Cold working increases the tensile strength and, simultaneously, the magnetic permeability of the wire. Only in its fully annealed condition can type 302 stainless steel be considered essentially nonmagnetic because in this state it exhibits the lowest permeability.

A unique industrial application for type 321 stainless steel wire fill material is in the hearth-style annealing furnaces used to process aluminum plate and cut sheet. With elevated temperatures of 1050°F (566°C) in some heating zones, many in-line wide-face cylinder brush rolls, slowly rotating in unison, support (by column strength) and convey the large plates and cut sheet through the multizone furnace. The brush rolls (rollers) are located inside the furnace and are subjected to elevated temperatures for years without deterioration. Replacement of the brush rolls occurs only when they are subjected to physical damage caused by an out-of-specification heavy plate curled edge hitting the brush roll directly and damaging the fill material beyond repair.

NONFERROUS METALLIC WIRE

Beryllium Copper—Nonsparking Wire

This fill material is a copper-based alloy, which is hardened by heat treatment, and is used in brushes to provide maximum protection against spark and explosion hazards. It provides good corrosion resistance, is nonmagnetic, and is recommended for nonsparking power-driven or manual brushing tools on many cleaning applications.

Nickel Silver

Nickel silver, a wire alloy predominated by nickel and elements of silver, is drawn into fine and very fine diameters (0.0025 to 0.0060 inches) with low tensile strengths. This nonoxidizing wire fill material finds many uses in jewelry, decorative silverware, the precious metals industry, and the electronics industry for mechanically cleaning and surface conditioning delicate parts.

Brass Wire

Brass wire is a composition of copper and forms of tin. It is relatively soft and low in tensile strength when compared to tempered carbon steel wire fill materials. The diameters of the wire are fine and very fine (0.0025 to 0.0060 inches), and the softness provides the advantage of nonscratching and/or nonmarking work characteristics. Applications range from sheet or epoxy mold cleaning, to luster finishing the surfaces of musical instruments, household and office fixtures, and a wide range of plumbing fixtures for residential and commercial buildings.

Brass-Coated Wire

Brass-coated wire is a product of the tire industry. It is used for tire beads and multistrand cord wire for steel radial belting. The brass coating (plating) on the carbon steel wire is necessary for the adherence of the compounded rubberized tire material. The wire is made in gauges (diameter sizes) similar to those of brush wire and is used on some applications where oxidation (rust) may be a problem for the high-tensile bright-finish carbon steel wire. The different physical characteristics of the brass-coated wire are a drawback in terms of performance criteria and life factors when compared to standard brush wires.

Zinc-Coated Wire

Zinc-coated wire is usually a hard-drawn and/or drawn and tempered wire that has a flash coating of zinc less than a half mil (0.0005 inches) in thickness. This wire fill material is generally used for consumer type products where the zinc affords an attractive optical appearance and some small degree of corrosion protection. However, with different physical characteristics compared to the standard high-tensile brush wire, the performance and life factors make this wire undesirable for the high-performance requirements demanded on many industrial applications today.

Plastic-Coated Wire

Plastic-coated wire is formed by taking tempered carbon steel wire and coating it with a thin (0.003 to 0.005 inches) plastic material. This will provide antirust qualities to the wire when it is used in wet or high-humidity environments. Since the plastic coating is placed on the steel wire at relatively low temperatures (approximately 250°F), the metallurgical qualities of the wire remain unchanged. Likewise, applications where high temperatures are involved will cause the plastic to soften and separate from the wire, affording little or no corrosion protection.

NONMETALLIC MATERIALS

Abrasive Monofilaments

Abrasive monofilaments are made up of two components: abrasive grains and nylon. During manufacture, the grains are homogeneously distributed and encapsulated throughout the nylon by weight, usually 20 to 45%. The diameters of the fill materials that are commercially available range from 0.008 inches to 0.080 inches, with many sizes in between. Other filament shapes, such as rectangular, provide a greater cross-sectional area, which contains more abrasive minerals and is thus stiffer in functional operation. Rectangular and other shapes are not crimped, similar to the round cross section ones available.

The abrasive grains most predominately used are silicon carbide and aluminum oxide. A few fill materials are available with alumina silicate as the abrasive. The grain sizes range between 1000 mesh (very fine) and 46 mesh (very coarse). The fine grains are associated with the smaller-diameter monofilaments, while the coarse grains are dispersed in the larger monofilament diameters.

Abrasive monofilaments can be used either wet or dry. Experience indicates some advantages of wet use when water, mineral oil, or water-soluble oil is used as a cooling medium to control temperatures. On dry applications, grease or pumice sticks may be used if elevated temperatures cause some smear or deterioration of the nylon in the fill material.

This category of cutting and grinding fill materials is the most rapidly growing in the industrial market. The abrasive grains are most suitable for working the very hard exotic metals and advanced composite industry component parts.

Clustered Abrasive Monofilaments

These fill materials are made by taking any number of single-strand abrasive monofilaments as described above, wrapping them with very thin yarns, and resin coating the entire assembly as a composite fill material. The abrasive grain and fill material stiffness is maximized to achieve superior grinding ability and product life in a wider range of applications.

Plastic-Coated Fiberglass

This fill material utilizes as a core strands of mildly abrasive fiberglass, encapsulated in a thin (0.003 to 0.010 inches) plastic, which provides the stiffness strength needed to make a usable composite fill material. It is a very effective wet scrubbing or surface cleaning material where very little abrasion is required. When used dry on applications without coolant or water, care must be exercised to avoid heat buildup, which will degrade the plastic in the fill material.

Nylon

Nylon is a high-molecular-weight polyamide fill material in monofilament form used for wet scrubbing and cleaning of material surfaces, as well as light dusting and sweeping in floor maintenance and care products. It is available in many different diameter sizes, usually in the fine (0.006 inches) to medium (0.030 inches) sizes. Wet scrubbing with temperatures above 120°F will create some loss of stiffness in the fill materials and will accelerate the degradation of the nylon.

Polypropylene

Polypropylene is a durable, abrasive-resistant, nonabsorbent, synthetic fill material used for wet scrubbing and cleaning of surfaces in elevated temperature environments (180°F to 210°F) and where strong detergents and solvents may be involved. The nonabsorbent characteristics cause it to retain stiffness, retarding brittleness and premature filament fatigue. Polypropylene is available in many diameter sizes, ranging from 0.006 inches to 0.080 inches, and is used on many industrial wet scrubbing and dry surface cleaning applications.

Polyester

Polyester is a smooth, pliable, synthetic fill material that exhibits good moisture resistance. It is used on specialized wet or dry industrial applications requiring fine wiping and light cleaning requirements. This material is also used in woven sheet form for use as buffs or combined with cotton fibers, in the buffing and polishing industry.

Aramid

Aramid is a very high strength-to-weight ratio synthetic material used in the advanced composite industry relative to aerospace and aircraft as a reinforcement component. Its high heat resistance, excellent fatigue resistance, and long-wearing qualities are useful in applications of mild cleaning in harsh, hot, and corrosive environments.

Carbon/Graphite

Carbon/graphite is a relatively new synthetic material used in the composite fabrication industry for its high strength light weight ratio and excellent reinforcing characteristics. As a fill material for industrial brushes, it exhibits high heat resistance, good fatigue characteristics, and very mild abrasive cleaning properties. Its noncontaminant and nonreactive makeup makes it a good choice of material on plastic and composite work pieces.

Vegetable Fibers

Tampico

Also known as Mexican fiber, tula, istel, and tampico hemp, tampico is obtained from the leaves of the agave plant, which grows wild in the semidesert higher elevations of Mexico. It is processed and sold in various grades, such as yellow, white, polished, sorted, and mixed.

Primary industrial uses are for hot wet scrubbing, cleaning, and on occasion for dry wiping and/or dusting. Because the fibers are porous (cellular), they absorb moisture (water) and provide an aggressive degree of cleaning. Tampico's lack of abrasion resistance on heavy industrial applications does cause rapid wear characteristics when compared to synthetic materials, although it can withstand higher operating temperatures than the synthetics.

Sisal

Sisal is obtained from the leaf of the agave sisalana plant in the same regions of Mexico as tampico. It is a stiffer material, usually produced from the butt ends of the long fibers. The porous nature of the material makes it an excellent choice for buffing wheels. It is then used with an abrasive compound for rapid and consistent cutting action.

Bass, Bassine, Bahia Bass

These three materials make up a closely related group of heavy, thick and stiff fibers from leaf stalks of the attalea plant, which grows in regions of Brazil, Africa, and India. They are primarily used for industrial and street sweeping applications, although some of the finer bass has been mixed with tampico fibers to provide greater stiffness on wet scrubbing applications.

Palmetto and Palmyra

Palmetto and palmyra are produced from palm tree leaf stalks and are similar in texture to the medium and fine bass fibers. They are products of Florida and India. The fibers find use in scrubbing brushes and coarse brooms. They are gradually being replaced by synthetic materials.

Cotton

Cotton is a soft, white, fibrous substance made into cloth sheeting and subsequently used as polishing and buffing materials. The porous structure and softness provide an excellent base to hold and maintain the cut and coloring abrasive compounds used in industry. Some buff materials are a combination of cotton and polyester (synthetic) fibers.

Linen

Linen is a fabric or thread made from flax plants and made into sheeting for polishing and buffing materials. It is used for special fine buffing applications because of its high quality. The material cost, when compared to cotton, is usually more because of the high quality.

3

Product Categories

POWER BRUSH TOOLS

Power brush tools used in industry are classified into seven major categories as determined by their geometric shapes and construction. Each category or shape relates to the adaptability of the product to the application work to be performed, such as deburring, edge contouring, cleaning, polishing, and surface conditioning, using the cutting and grinding fill materials described in Chapter 2. These major categories are as follows: (1) wheel/radial, (2) cup, (3) disc, (4) end, (5) twisted-in wire, (6) cylinder (wide-face), and (7) strip brushes.

For a better understanding of the power-brush major categories, each will be divided into the important subparts and outlined.

Wheel/Radial Brush Tools

Wheel brushes are circular in shape with uniform distribution of the fill materials extending radially from the center core to the periphery.

Concentricity and balance are important design criteria because of the high speeds and centrifugal forces that this type of brush is subjected to during operation.

Wheel brushes are the most popular and most used in industry today, ranging in outside diameters from $\frac{1}{2}$ inch to 24 inches and widths from $\frac{1}{8}$ inch to 3 inches. From a practical standpoint, when the width of the brush exceeds the outside diameter, the category then becomes a cylinder brush.

As pictured in Figures 3.1 and 3.2, brushes of the twist-knot construction type are almost exclusively made utilizing straight metallic wire fill materials, while the crimped construction type encompasses both metallic and man-made synthetic fill materials. The natural fibers are used in their normal straight form. They are too brittle to crimp or twist and therefore, used in wheel brush form similar to Figure 3.3 with or without stitching.

Wheel brushes are available in short, medium, long, and extra long fill material trim lengths. Trim length is defined as the length of the fill material extending from the core side plates to the periphery or outside diameter of the brush.

Trim length plays an important part in the performance and success of each product application and can be best outlined as follows:

Short Trim: Less Than 1 Inch in Length

In wheel brushes, short trim is characterized by densely filled (maximum fill) material, stiffness, and minimum flexibility. These brushes provide the maximum cutting or grinding action on the work piece by utilizing the tips of the fill materials. This is typical for both twist-knot or crimped-type constructions.

They are used extensively for the removal of burrs from gears in the automotive, agricultural machinery, and power transmission industries. These short-trim brushes impart the proper radii and adequately blend the edges of gears while at the same time removing all the minute fragments of metal that infrequently are left when other techniques of burr removal are previously employed. One distinct advantage of this type of brush is its capability of removing the burr from an edge without changing the dimensional characteristics of the two intersecting sur-

Figure 3.1 Twist-knot type wire.

Figure 3.2 Crimped type, abrasive.

surface conditioning, and the brushing of uneven surfaces for scale, rust, dirt, or loose paint removal.

Extra Long Trim: From 4 Inches and Over

The longer the trim length of a wheel brush, the more flexible the fill material becomes, thereby providing some unique features necessary for specific applications. Large-diameter wire wheel brushes with extra long trim are less susceptible to heat generation when applied to iron and steel parts that have temperatures in excess of 300°F. Cleaning welds on pipeline applications is an example of this.

Also, large-diameter abrasive monofilament wheel brushes having extra long trim and rotating slowly do provide a side-wiping action of the fill materials (including the abrasive grain) that removes burrs and contours edges of deep-pocketed, unusual geometrically shaped aerostructure parts. The term "mush brushing" has been coined for similar applications where the brush fill materials are "buried" into the part in order to contact many areas at one time. The buffing industry has used this technique for many years using a similar term—"mush buffing."

Centerless Radial Brushes

Centerless brushes fit standard centerless grinders. These large radial brushes have a wire fill (0.005 to 0.014 inches in diameter) or a fill of abrasive monofilaments, treated tampico, or cord, with a cut or color abrasive compound. When the fill is wire or abrasive monofilament, these brushes remove grinding burrs and produce a minimum edge blend 0.005 inches. Finishing work can be done either dry or, preferably, with normal grinding coolant.

Fine wire fill or abrasive monofilament brushes, working with conventional grinding coolant, remove feather-grinding burrs and improve surface finishes on such parts as control valves for automatic transmissions. In such applications they can improve the finish from 8 to 4 rms (foot mean square). If the fill is treated tampico or cord, applying a grease-based polishing compound to the wheel face will give finishes in the range of 2 to 8 rms.

It is important to remember the following facts about centerless brushes:

1. They will not remove metal from a cylindrical surface. Parts must therefore be ground to size before brushing.

2. Greatest improvements in finish (for unbrushed parts) are in the 25 to 30 rms range. A part with a 25 rms finish can be brushed rapidly to obtain a 10 to 15 rms finish; a part with a 10 rms finish will have a 4 to 6 rms finish after brushing, but the time cycle will be much longer.

3. Centerless brushing follows centerless grinding principles, with the following exceptions: Accuracy in pressure and adjustment is not critical with brushes, and a close-tolerance grinding machine is no longer needed to obtain the required finishes. In fact, an older, lower-precision grinder would be an ideal tool to use with centerless brushes.

Cup Brush Tools

This category contains a number of different types and variations of the cup brushes wherein the uniform distribution of the fill materials is contained in a backing and parallel or slightly angled to the brush driving spindle (shank), as in Figures 3.4, 3.5, and 3.6. Equally important are the concentricity and balance design criteria because of the centrifugal forces generated by the speeds and application requirements during operations.

The most popular industrial use of the wire cup brush tools has been on portable hand-held power-driven air or electric tools where it is more economical and convenient to bring the power brush to the larger work pieces. In addition, some brushes of this type are used on automatic and semiautomatic industrial machinery and pipeline equipment. Recently, because of the success of the abrasive monofilaments, extensive use is now made of abrasive cup brush tools on numerical controlled (NC) and computer numerical controlled (CNC) machine tools. The abrasive finishing operation is integrated into the machine after the machining operations (see Figure 3.7). Cup brushes range in outside diameters from $\frac{1}{2}$ inch to approximately 8 inches and utilize twist-knot and crimped construction in metallic, man-made synthetic and natural fiber fill materials.

Figure 3.4   Twist-knot type cup brush.

Figure 3.5   Crimped type cup brush.

Figure 3.6   Cup brush, crimped type with shank (stem).

Figure 3.7   NC and CNC cup brush tools.

The trim lengths of the fill materials are narrower in scope than the wheel brushes, basically having short and medium lengths.

Short Trim

Short trim here is defined as less than 1 inch in length. It is used for stiffness, aggressiveness and minimum flexibility, using the tips of the fill materials primarily. Some typical applications include pipe end cleaning, weld scale removal, deburring gears, machined steel fabrications, heavy paint and rust removal, hydraulic valve body parts, and most flat surface deburring and finishing work.

Medium Trim

The medium trim, ranging from 1 inch to $2\frac{1}{2}$ inches, is chosen for various degrees of flexibility to follow uneven contours and unusual geometric shapes using both the tips of the fill materials as well as the sides on some applications. It is more desirable when using metallic fill materials to focus on using the tips for the most effective cutting and impact against the work to deburr, edge contour and surface condition. The sides of the metallic fill materials are smooth and have no cutting characteristics.

With nonmetallic fill materials, such as abrasive nylon monofilaments, using the tips and/or the sides is very effective as the abrasive grains are uniformly exposed at the tips as well as the sides. Using the side action of the abrasive provides a ''draw filing'' effect on the work piece, especially when edge contouring or surface conditioning.

Long Trim

The long trim is from $2\frac{3}{4}$ inches to 4 inches. Almost all are abrasive monofilament tools because of the cutting action on the sides of the filaments. The more flare of the filaments, the more abrasive contact is made on the work piece and more work is produced per unit of time.

Disc Brush Tools

Disc brush tools are usually flat backed with a center hole or adapter for mounting on a drive spindle and having the fill materials extend out perpendicular to the backing material (Figures 3.8 and 3.9).

Figure 3.8   12-inch disc brush tool with abrasive nylon strips.

Figure 3.9   12-inch disc brush tool, face down.

Disc diameters range from 4 inches to 48 inches, and the trim lengths of fill material range from 1 inch to 8 inches. By far, the abrasive nylon monofilaments in rectangular and/or round form is the most popular material for deburring, edge contouring and surface conditioning, because of the abrasive cutting action of the tips and sides of the filaments.

Operational speeds are in the area of 2000 surface feet per minute (sfpm), and the tool provides a circular rotational motion to achieve a 360 degree contact with all areas of the work parts being finished. The rotational motion can be in either direction or made orbital by the machine tool.

Disc tools have inherent variable speeds built into the filaments, based on their specific diameter location in the flat backing material. For example, a 12-inch outside diameter tool having a multiplicity of filaments to a 6-inch inside diameter and rotating at 600 rpm will produce 1885 sfpm at the 12-inch diameter and 943 sfpm at the 6-inch diameter with incremental variations in between. In order to achieve an operational speed of 2000 sfpm, take the average diameter of fill material at 9 inches and the calculation for rpm comes out at 849 revolutions per minute, which is the spindle speed.

Applications include the deburring of large outer case parts with scallops or cutouts, hydraulic valve bodies, machined transmission cases, flat sheets and plates, and double disc ground parts.

End Brush Tools

End brush tools are usually small in diameter (up to 1 inch in body width) with a mounting shank (spindle) being an integral part of the body. The fill material length parallels the shank and body, extending outward from the inside of the body, thereby providing various degrees of trim length, as in Figure 3.10.

In operation, usually at higher speeds over 10,000 revolutions per minute (rpm), the fill materials will flare from the centrifugal forces generated and become larger in diameter, covering more work area. On the contrary, when used at slower drill press speeds of around 2000 rpm no flare occurs and a ''spot facing'' effect is produced.

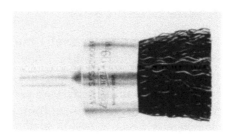

Figure 3.10 End brush tools.

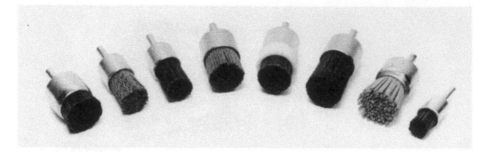

Figure 3.11   NC and CNC end and spot face tools.

There are many types of end brushes. Those with solid faces, hollow center, with pilots, pencil construction, etc. may be used on applications like mold cleaning, fabrication weld and pipe cleaning, spot facing, and unusually shaped geometric areas hard to get at. Typical examples of some end or spot facing abrasive tools used on numerical controlled (NC) and computer numerical controlled (CNC) machine tools are shown in Figure 3.11. Almost all of these products are made from abrasive nylon monofilaments with rectangular and round shapes and abrasive mineral sizes ranging between 80 grit and 600 grit mesh in silicon carbide, aluminum oxide, and polycrystalline diamond.

There is such a wide variety of fill material trim lengths that it is difficult to generalize on a particular size. Depending on the makeup of the fill materials, metallic or nonmetallic, and on the rotational speeds used, the shorter length material (1 inch or less) is found in the metallic range of brushes. Nonmetallic and abrasive fill materials generally range from 1 inch to 3 inches, as the material weight is less than metallics and the needed rotational speeds are less.

Twisted-In Wire Brush Tools

Twisted-in wire brush tools are generally referred to as tube or hole cleaning brushes or side action brushes. They are used to deburr, edge contour, surface condition, and clean internal openings, threaded holes, slots, cross holes, etc. The fill materials are perpendicular to the retaining stem wire or are fashioned in helical form but still assume perpendicularity to the stem wire, as in Figure 3.12.

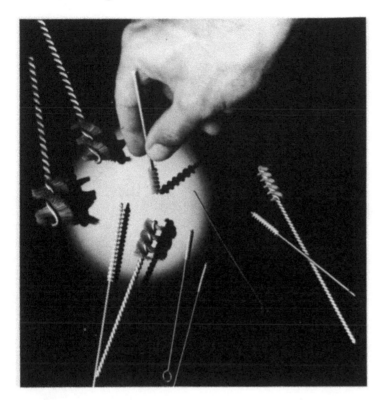

Figure 3.12   Microabrasive tools.

The tube or hole cleaning brushes are operated at drill press or hand drill speeds (approximately 2000 revolutions per minute or less) and need to be inserted into the holes very carefully to avoid bending or damaging the brush stem.

In some special cases it is necessary to insert the brush into the hole in a static (nonrotation) mode and then start the power rotation for the brushing operation. Care must be exercised to stop the rotation of the brush *before* removing the brush from the hole, as the centrifugal forces of the rotation may bend the stem wire or securing device of the fill materials, which are usually silicon carbide, aluminum oxide, alumina silicate, and, in some cases, polycrystalline diamond.

Brush size ranges vary infinitely in diameters, from the micro for cleaning aerospace, medical, computer, and hydraulic components to

Figure 3.13   Macroabrasive tools.

the macro size of 8 inches for honing and finishing automotive and truck engine cylinder bores, as in Figure 3.13.

Brush tools of this type are extremely important to NC and CNC machine tools with tool change magazines. Parts that are turned, milled, drilled, and ground can be deburred, radiused, and finished with a simple tool change programmed into the machine cycle and the piece parts finished in one total machine operation. This eliminates the need to finish the work piece off-line on a separate operation or at the hand deburr bench.

Cylinder Brush Tools

The best definition of a cylinder brush is a radial/wheel brush where the face width exceeds the outside diameter; Figure 3.14 illustrates this brush type.

Figure 3.14   Cylinder brush tools.

Probably 90% or more of all wide-face brushes are custom made to suit the particular requirements of the user. They are made in diameters ranging from 2 to 30 inches and in lengths up to 25 feet. Every conceivable type of fill material is used, including all the popular synthetics, hairs, natural fibers, and abrasive monofilaments and metallic wires.

Although the customer must provide specifications of overall design and dimensional characteristics, a specialist in brush-making technology will provide the detailed specifications relating to the kind of fill density, trim length, and type of construction desired. The brush manufacturer should also determine whether the brush should be made with wheel sections mounted on a common arbor or with a continuous strip helically wound into convulsions of sufficient number to produce the proper face width. The manufacturer should suggest whether the brush is to be used wet or dry and what pressure and speeds need to be considered.

It is not uncommon for worn brushes, mounted on permanent arbors, to be returned to the manufacturer for replacement. The brush manufacturer removes the worn brush parts from the arbor, replaces, trims, and dynamically balances the brush assembly, and then returns it to the customer. For some less sophisticated applications, a helically wound brush element or a group of sectional brushes may be ordered by the customers for self mounting on their own arbors.

A combination of these two concepts, incorporating the advantages of each, is the unitized section or disposable unit. For this brushing method, the brush manufacturer fabricates wide-face brushes on a thin-walled steel tube, which contains inserts that reduce the tubing to the customer's arbor size. These finished units are forwarded to the customer, who merely slides them onto his arbor and locks them into position. Since the worn brush elements may be thrown away, the cost and time required to ship arbors back and forth between manufacturer and customer are eliminated. Minimal spare parts inventories are required when this system is used. The disposable unit brush is used extensively in the steel industry for scrubbing steel that is processed on galvanizing lines, electrolytic tin-plate lines, anneal/pickle lines, cold reduction cleaning lines, paint coating lines, and roller levelers. In some of these applications, the brushes are the flow-through variety to facilitate a continuous flow of water or mild alkaline solution through the brush and arbors.

The circulation of water, which is emitted through the face of the flow-through brush, serves a number of purposes over and above dissipating the heat generated in brushing. It also serves to dissolve the dirts and oils on the strip being cleaned and, most important, it flushes the brush face free of dirt.

The flow-through principle is not confined to throw-away can brushes, but is also frequently found on the refillable, permanent-arbor types. Similar scrub or washer lines are found in aluminum, copper, and brass mills and in some instances in mills for sheet-glass products.

Precision-ground metallic and abrasive monofilament brushes weighing hundreds of pounds each are used to control the aluminum oxide buildup on the steel work rolls of continuous and reversing hot mills in the aluminum industry.

Special stainless steel wire filled brushes, covering stainless steel conveyor rolls, are used directly inside hearth furnaces to support and convey aluminum and alloy plate through the furnace for attainment of critical metallurgical properties and smooth, nonscuffed surfaces of the plate.

There are more applications for wide-face brushes used dry than there are for those used wet or with a slurry. The textile industry alone uses literally hundreds of different types of brushes, and tanneries use brushes for seasoning, dusting, buffing, polishing, and oiling. The paper processing industry requires brushes for dampening, for buffing clay-coated stock to attain high-luster finishes, for coating and dusting, and for cleaning 4-denier wire mesh. The lumber and furniture industries use brush tools to sand, polish, and grain wood products to develop unique and contemporary surface finishes. Other applications for wide-face brushes include thousands of miscellaneous operations in almost every type of industry.

Strip Brush Tools

One of the most versatile brushes available is the strip brush, which can be customized by either the manufacturer or the user, depending on the purpose of the brush (Figure 3.15). For example, virtually any strip length, height, channel retaining material or channel size desired can be obtained. The choice of fill materials includes the entire spectrum of synthetic, vegetable fiber, hair, abrasive monofilaments, and metallic wire filaments.

Applications of strip brushes are as limitless as the variety of brushes available. Strip brushes are used extensively as splash curtains or curtain walls on machine tools and ovens: as lag brushes, as static electricity eliminators and backups for coated abrasive wheels, for dusting rubber goods such as tires, and in many vegetable and fruit cleaning operations. One of the widest uses of strips is for assembly on rotary hubs or extruded mountings to form rotary brushes for the cleaning of conveyor belts and screens and the spreading of various materials. The strips are mounted either straight, parallel to the axis of the brush, or helically around the axis to ensure continuous contact with the work. Because of the space between the strips, the brush does not load up and its flexibility permits it to conform to the irregular surfaces on which it is used.

The strip brush tools utilizing the rectangular/flat abrasive nylon monofilaments with extra long lengths are used extensively in the aerostructurers industry on robots to remove burrs and contour edges by "draw filing" burrs away using the sides of the long (8 inches) filaments (Figure 3.16). Because of the slow rotational speeds, the rectangular or round filaments are continuously wiping across the sharp edges with 3 or 4 inches of filament lengths, grinding away the burrs and generating edge radii.

Elastomer-Encapsulated Brushes

This classification of power brushes crosses all seven major categories regardless of brush geometric shape and construction, as shown in Figures 3.17 and 3.18. The addition of a resilient, tough, thermoplastic elastomer resin to wire fill material brushes orientates, stiffens, and exposes only the sharp wire tips (points) at the periphery or working area of the brush. Different degrees of stiffness are obtained by changing the chemical elements in the formulation of the resin.

Figure 3.15   Strip-wire filaments.

Figure 3.16   Rectangular abrasive strip brush tool.

Figure 3.17   Elastomer-encapsulated brushes.

Figure 3.18   Elastomer-encapsulated narrow-face brush.

Performance of the elastomer-encapsulated products usually generates twice the work per unit of time and five to eight times the average life of a conventional short-trim wire brush on comparable applications and operating conditions. These brushes are excellent for automated equipment with a high volume of parts processed at a short time cycle with very few pressure adjustments needed.

However, elastomer-encapsulated brushes are not a "cure-all" by any means as a replacement to other standard wire brushes constructions, because of the limitation of wire flexibility needed for many applications and the possible rapid degradation from elevated temperatures (over 250°F) from the heat generated by deburring the work pieces and/or parts.

POWER BUFFS

Power buffs are generally categorized as wheels and subdivided into the following seven major categories: (1) full disc, (2) packed, (3) pierced, (4) bias, (5) ventilated, (6) finger, and (7) goblet.

To understand the power buff major categories, each will be described and outlined by their subpart.

Full Disc Buffs

Full disc buffs (Figure 3.19) are manufactured from cotton or cotton and polyester sheet materials and assembled into a number of plies or layers to make buff sections of various thickness. Some sections are sewed around the arbor (center) hole to provide additional strength. The surface area may be sewn with different patterns, with the one shown having concentric circles. Outside diameters as large as 24 inches are not uncommon today.

Packed Buffs

Packed buffs (Figure 3.20) are similar to full disc buffs except that smaller diameter plies of material are put in between the larger ones to provide a loose or open-face buff for light-duty applications. Stitching is necessary around the arbor hole for extra strength because the surface

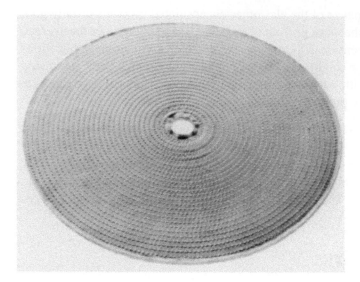

Figure 3.19   Full disk buff.

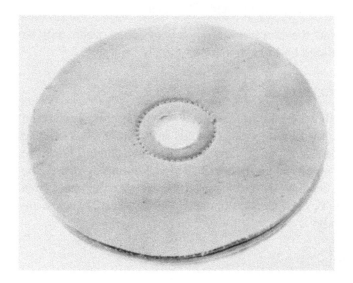

Figure 3.20   Packed buff.

areas are not sewn to provide the open face characteristic of this buff. Outside diameters usually are as large as 14 inches.

Pierced Buffs

Pierced buffs (Figure 3.21) utilize recycled cloth materials with less critical physical specifications. They are assembled as a collage and sewn together with a full disc-type stitching pattern and sold at very economical cost.

The pierced buffs are primarily used on coarse cut-down buffing applications.

Bias Buffs

Bias buffs (Figure 3.22) are the most popular and universally used for both cut and color buffing on semiautomatic machines and on fully automated manufacturing systems. The bias of the material minimizes unraveling, plus utilizing the metal lacing for the center hole provides dimensional concentricity and a strong positive lock for the material used in the ventilated or nonventilated mode. The geometric pattern and shape of the material at the buff face (sometimes referred to as pucker) generates an overlapping effect when applied to the work. This effect attacks straight-line scratches in the work from a slight angle rather than parallel to it. Buffing wheel diameters are made as large as 24 inches.

Ventilated Buffs

The ventilated buff is shown in Figure 3.23. Ventilation and/or air cooling, while in operation, is an important feature to buff life and performance and is similar to that being used on some power-driven brushes. The center plate holes and slots draw ambient air created by the rotational speed of the buff and force it in between and through the layers or plies of material to the periphery of the buff. Heat buildup is reduced, and the small cavities developed by the air passing through grips additional cut or color compounds, thereby providing more abrasive against the work pieces. The ventilation feature can be used with most buff materials, including cotton/polyester buffs, which are more heat sensitive.

Figure 3.21   Pierced buff.

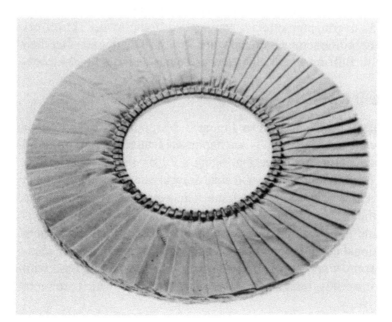

Figure 3.22   Bias buff.

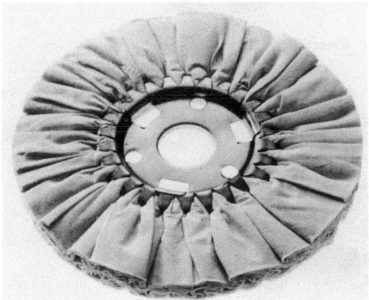

Figure 3.23 Ventilated buffs.

Figure 3.24   Finger buff.

Figure 3.25   Goblet buff.

Finger Buffs

This style of buff is made from folded cloth ranging from very soft cotton sheeting to the course natural fiber sisal. Depending to the degree of stiffness required, various amounts of stitching can be added to reduce flexibility, making for an aggressive cutting buff. The finger buff shown in Figure 3.24 uses the metal lacing for the center hole and is designed for ventilated operation.

Goblet Buffs

Goblet buffs (Figure 3.25) are 6 inches or less in outside diameter with thickness to 4 inches, used to color buff internal surfaces, contours, and unusual inside geometric shapes. The construction design requires the threads of cloth to extend outward from the center to accept and hold the color compounds on the spherically contoured surfaces. The small center hole accepts screw points, tapered rods, or spindles to hold and rotate the goblet buff during operation.

4

Construction Design Criteria

INTRODUCTION

The application of power-driven brushes and buffs to deburr, edge contour, blend, radius, surface improve, condition, clean, and perform other functions is best defined as a state-of-the-art technology rather than a science, as exemplified by metal turning, milling, drilling, or grinding where under given conditions of data, very predictable and consistent results in metal removal rates, surface roughness characteristics, and the like can be generated.

Good information and technical data are continually being developed by the leading brush and buff manufacturers with the help of many product users. High-volume continual production of standard parts provides the most reliable data gathered on accurate performances and costs generated by specific brush and buff tools.

Included in Chapter 5 is an overview devoted entirely to the application of brushes and buffs with technical operating data and some cost information. But first, let us review brush selection and criteria involved.

BRUSH FILL MATERIAL SELECTION

The choice of fill materials for a power-driven brush tool should take into consideration a number of the following physical characteristics:

> Basic materials—man-made or natural
>
> Elasticity—resilient, ability to withstand shock
>
> Diameter—including tolerances
>
> Formation—crimped, knot, straight, multistrand crimped
>
> Trim length—material length, retaining member to outside diameter
>
> Rotational speed—centrifugal forces generated
>
> Density—points per square inch at periphery

Metallic Filaments

Every power-driven brush tool constructed of metallic filaments uses one of the following basic forms:

Crimped Wire Filaments

Many types of crimped wires are used in brushes to mainly provide a uniform density of wire distribution. The best crimping results are obtained when the wire is displaced over multiple planes. Although there are a number of crimping techniques that will produce multiple-plane displacement, the most common is similar to the dual-plane crimp configuration produced, being approximately 90 degrees apart. The wavelength of the crimps, sometimes called frequency, should be uniform over the entire length of wire. The displacement of the wire, usually referred to as amplitude, should be held to a very close tolerance. Uniformity and consistency of frequency and amplitude are essential in high-quality brush performance.

Multistrand Crimped Wire

This is basically straight wire that is stranded together by twisting into cable form similar to tire cord or tire bead wire. After stranding, the multiple strands of fine wire (0.010 inch diameter or smaller) are crimped using the same process as with single-strand wire. The effect of multistranding is to provide additional fatigue strength and stiffness to the fine wire while maintaining the smooth uniform cutting action on the work pieces.

Knot or Twisted Tuft

This construction method is the most desirable for heavy, aggressive, high-impact application work because the straight wires are twisted together in small cable-formed tufts. This provides each wire filament additional support and stiffness from a multitude of adjacent filaments. The series of knots are held in a circular form by a center disc, ring, or retaining member to form the basis of the knot or twisted tuft power-driven brush.

Variations in degrees of the cable-formed helix can provide more or less stiffness and/or flexibility to the brush. Multiple rows of tufts mounted on a common center disc can maximize the total wire and tuft count, providing an extremely stiff, aggressive brush tool.

Straight Wires

Brushes manufactured from straight wires, that are not crimped or twisted in knot form generate a high degree of flexibility because the individual wire strands are unsupportive of each other and essentially stand alone when applied to the work piece. The common manufacturing technique used is referred to as "staple set" or "punch fill." A series of patterned holes is drilled into a backing material made of wood, plastic, or composites. Small bundles of straight wire are automatically hairpined and driven into the holes using a staple wire, which imbeds into the backing material and locks the wire bundles in place. Brushes of this construction have a limitation on operating speeds, much less than with the other types of construction.

Nonmetallic Filaments

A substantial portion of the applications requiring very fine finishes of 4 to 32 rms simply cannot be accomplished with wire filled brushes. Similarly, applications requiring a reduction of stress concentrations can, in many instances, be best accomplished by means of brushes constructed of nonmetallic fill material. Three primary fill materials for this type of work are abrasive nylon monofilaments, tampico, and string and cord. The tampico and the string and cord require an abrasive compound (liquid or solid) to generate the fine finishes and to blend out scratches, thereby reducing stress concentration points on parts.

Abrasive Monofilaments

The abrasive monofilaments currently used in brush tool forms have advanced the state of the art of brushing more than any other fill material in the Brush industry. This is not yet a science. However, as more data and technical information are generated by users and by research and development, the science will become similar to that of the grinding industry. There are many similarities to grinding technologies, with the difference that the abrasives, securely held in nylon-based fill material, are "flexible" when applied to the work and not rigidly held as with hard grinding wheels.

Basically the abrasive monofilaments are comprised of two elements, *abrasive grains* and *nylon*. As the nylon is extruded into various shapes (round, rectangular, oval, etc.), precise amounts of abrasive grains are added to the nylon, resulting in a finished product having a specific amount of abrasive grains homogeneously distributed throughout the filaments.

The filaments used in brush tools can be straight, crimped, or multistranded or yarn wrapped. Minerals available today include silicon carbide, aluminum oxide, silicate, polycrystalline diamond, and others. Loading (weight of grain versus nylon) ranges between 10% and 45%, with the heavier loadings in the larger filament diameters up to 0.080 inches and the rectangular cross sections of 0.045 by 0.090 inches. The smallest diameters are approximately 0.008 inches.

Recently a new flexible abrasive material has added a dramatic dimension to the finishing tool market once dominated by round-shaped abrasive filament materials, nonwoven abrasives, buffs and compound, and some coated abrasive products. *Rectangular* or flat-shaped abrasive and nylon monofilaments used in the same tool form with the same sized abrasive mineral generate a "line" contact at the work surface of the piece parts. When the monofilaments are shaped rectangularly (0.090 inches wide and 0.045 inches thick), the difference between "line" and "point" contact by round filaments translates into approximately a factor of *plus eight times* abrasive-on-surface cutting for the rectangular-shaped material. The round monofilament initially starts cutting with a point contact and eventually wears itself to a level of flatness during the normal life of the tool. Continual ongoing changes in the round filament flatness during operation of the tool do not lend itself to the consistent, repeatable, part-to-part quality results of the deburring and edge contouring as performed by the rectangular flexible abrasive grinding tools. The initial cutting of the rectangular filaments by line contact remains the same throughout the life of the tool, with worn minerals being removed by new ones.

Second to the shape change is the elimination of "crimping" or producing a sine waveform in the rectangular abrasive and nylon monofilament. Keeping the line contact material in straight (uncrimped) form continuously maximizes the abrasive mineral contact at the work surface. This produces a continuous draw-filing or grinding action on the parts. The larger cross-sectional area permits upward of 45% loading of abrasive mineral without a negative effect on the overall filament strength. In terms of industrial usage, in flexible abrasive finishing tool forms, there is a 50% gain in mineral content using the new rectangular-shaped material.

The brush tools manufactured with this material can be used either wet or dry. Coolants are generally recommended to reduce heat, remove swarf, and provide better tool life. In the cases of dry applications, grease sticks—used sparingly—will provide lubricity to the application and reduce heat buildup, which may otherwise have a deteriorative effect on the fill materials.

Tools made with this material can be used to deburr, edge contour, radius, surface finish, and clean the very hard superalloys, steels, titanium, aluminum, carbides, composites, advanced composites, and some ceramics.

Tampico Fibers

The use of tampico fiber brush tools with cutting and polishing compounds has diminished in recent years, with many industrial applications replaced by the abrasive monofilament products. The tampico fiber brush tools, to be effective for deburring, edge contouring, and surface conditioning, needs a semitack resin applied throughout the fibers. This treatment will hold and apply "cut" and "color" abrasive compounds, which perform the work of metal removal and blending, just as with buffs in the buff industry. The abrasive compounds used can be either liquid or solid bar types. The rationale for gravitating toward the abrasive monofilament products has included the lower end-of-service costs of abrasive monofilament products, the excessive throw-off and waste of compounds during the brushing operations with tampico fibers, and the necessity for putting the piece parts through an added intensive washing and cleaning operation to remove the cutting compounds and the resin residue deposited by the tampico fiber brushes. This cleaning operation is not necessary with the flexible abrasive monofilament products.

String and Cord Fibers

The use of small-diameter soft cotton yarn (string) and large-diameter cotton, linen, or hemp (cord) with cut and color abrasive compounds has diminished considerably over the years, having been replaced by various buffs with similar materials, and the finer mild abrasive nylon materials. String and cord fiber tools require liquid and/or solid bar compounds to do the work, similar to the tampico fiber tools. Similar washing and cleaning operations are required on piece parts to remove the compounds and resins.

ELASTOMER-ENCAPSULATED BRUSH TOOLS

This special category of brush tools is unique in that a resilient chemical additive, similar to polyurethane, is molded throughout, in between and surrounding all filaments except at the exposed face or filament tips that contact the piece parts during the work cycle.

The resilient chemical additive can be applied to all the major categories of brushes—wheel and radial, cup, disc, end, twist-in wire, cylinder, and strip—to severely restrict the amount of filament flexing and maximizing the cutting action of the filaments.

During normal operation of the encapsulated brush, the elastomer (chemical additive) wears back at the same rate as the wire tips when they become dull and break off in $\frac{1}{8}$ inch lengths.

Three grades of elastomer hardness are normally available, ranging from soft to medium to hard, varying between 70 and 90 on the durometer A scale.

Having very limited flexibility, this product is well suited for applications where filament penetration is not necessary, such as flat surfaces and light burrs. The elastomer, being a chemical combination, is susceptible to deterioration from elevated temperatures when in use. Continual operating temperatures over approximately 250°F will degrade the material, causing it to prematurely dissipate from the brush tool and thus causing the wire filaments to lose stiffness. In some respects, the appearance of this tool is very similar to a hard grinding wheel with wire tips protruding at the working face instead of abrasive mineral.

POLISHING AND BUFFING

Polishing

Polishing is a primary mechanical finishing operation that uses selective abrasive grit minerals brought to and held against the work by semiflexible support systems, such as endless abrasive belts, abrasive flap wheels, and soft flexible grinding wheels.

Conventional hard grinding wheels are sometimes used to quickly bring the piece parts to close approximate size and tolerance that will be suitable for the start of the intermediate polishing operations. The heavy burr removal, edge contouring, and surface conditioning are performed by the polishing tools available in preparation for the buffing operation, which provides the smooth refined and high luster finishes with cloth and natural fiber buffing wheels in conjunction with abrasive compounds.

A typical flexible coated abrasive flap wheel is shown in Figure 4.1, deburring the edges and surface conditioning a three-bladed marine propeller part. Various abrasive mineral sizes are available in silicon carbide and/or aluminum oxide to suit the final requirements. As with the case of buffs, these polishing wheels can be contoured to different shapes that are suitable to the part geometries.

Buffing

Buffing is also a primary mechanical finishing operation using cloth or natural fiber as backing and holding materials for the sized abrasive grit minerals held together by an assortment of liquid or solid binders.

Standard buffing techniques consist of a dual-stage operating procedure of *cutting* and *coloring*. The cutting stage usually follows the polishing operations and provides an additional refinement to the surface by removing scratches, marks, and oxidation and other surface blemishes, and it preconditions the work for the final coloring (smoothing) stage.

Coloring reduces the cut surface to a high-shine, mirror-like, reflective radiance using very fine abrasive grain or powder compounds.

Typical cut and color buff construction and hardware components have been described in Chapter 3.

Buff-use abrasive compounds are readily available in two basic forms, the liquid for high-volume, high-production requirements and the solid bar form for specialized applications. The solid bar form is more applicable in the color buffing stages of finishing.

Some of the most common abrasive mineral grains or powders used with buffs for nonferrous metals are Tripoli, soft silica, and combinations of the two. Being soft in nature, these compounds are more forgiving than the aluminum oxide mineral grains or powders used on most ferrous metals, except for the use of chromium oxides, which are used primarily on the nickel/chromium stainless steel work parts.

Heat-sensitive work such as on plastics and synthetics would be

Figure 4.1   Flexible coated abrasive flap wheel.

better suited by a very fine pumice and water slurry mixture to dissipate heat buildup in the work during buffing.

Binders are necessary to hold the abrasive minerals in suspension with liquid compounds and also to form the permanent shape of the solid bar compounds. Some of the more common binder materials are waxes, stearic acids, tallows, petroleum greases, and adhesives.

5
Applications, Technical Data, and Costs

In order to better understand the use and operational features of brush and buff tools we must first look at, study, and understand the safety requirements and recommendations for all persons involved in and associated with their uses.

Material removal operations include deburring, removal of unwanted material from machining operations, edge contouring, generating edge radii or edge blends on sharp corners, and surface conditioning, refining, or texturing surface areas of material.

Metal and material pieces, slivers, burrs, fragments, rust, dirt, debris, and other particulate matter are removed from the piece parts being worked on and are propelled with a force provided by the rotating brush and/or buff tools during the operation. This removal continues until the machine or tool is in a zero mechanical state of operation.

STANDARD SAFETY REQUIREMENTS

For most industrial operations there are standard safety requirements for dust or debris collection systems to evacuate and collect the waste materials being removed for the safety of the personnel and equipment and for good housekeeping practices. Good operating practices, when using brush or buff tools, should include the following safety requirements and recommendations:

Operator Preparation

All power brushes and buffs, like other rotating cutting tools, demand that certain operating precautions be observed to assure operator and work area safety. All operators must read and understand safety information thoroughly and completely before using the products.

Inspection

Products should be carefully checked when removed from original carton. Do not use if rusted or damaged.

Storage and Handling

Store products in original boxes. Wire brushes should not be exposed to heat, high humidity, acids, or fumes or liquids that can result in deterioration of the wire filaments and, subsequently, premature failure of the wires. Also, check for distortion of brush fill that can cause imbalance and excessive vibration when brush is run. Do not allow foreign material to accumulate in brush face.

Machine Condition

Proper maintenance of machines is essential to keep them in safe operating condition. Special operating instructions furnished with a machine should be closely followed. Hoods and safety guards must be kept in place at all times. Use an adequate spindle diameter for the products—do not use products larger than the machine was designed for. Machines should have sufficient power to maintain rated spindle speeds.

Mounting

Inspect for rust, oxidation, and other damage. Do not use the product if not in good condition. Check spindle speed (rpm). Do not mount and operate if spindle rpm exceeds *maximum safe free speed* (MSFS) for which product is rated. The arbor hole and spindle diameter should be the same for free fit. The spindle length should be sufficient to permit a full nut mounting. The direction of spindle nut thread should be in such relation to the direction of rotation that the nut will tend to tighten as the spindle revolves. When flanges are used, they should be identical in size and radial bearing surface to avoid cross-bending pressure.

Work Rest

On single-end or double-end pedestal machines, work rests of rigid and adjustable design should be used to support the work piece. Adjust the work rest for a maximum opening of ⅛ inch to the wheel face. This will prevent the work from being forced between the wheel and rest. The work rest should be adjusted only when the wheel is not in motion.

Speed

Maximum safe free speed (MSFS) is the maximum rpm at which the product should be operated with no work applied (spinning free). It is not the recommended operating speed. The application determines the recommended operating speed, which should never exceed the MSFS ratings marked on the product and/or shown in a catalog. Periodic speed checks of the spindle are the responsibility of the operator and user.

Protective Equipment

The potential of serious injury exists for both the operator and others in the work area (possibly 50 feet or more from the operation). To protect against this hazard, before rotating the product, during rotation, and until rotation stops, operators and others in the area must wear *safety goggles* or *full face shields worn over safety glasses with side shields.* Comply with the requirements of ANSI Z87.1-1979, Occupational Eye and Face Protection.

Appropriate protective clothing and equipment (such as gloves, respirators, etc.) shall be used where there is probability of injury that can be prevented by such clothing or equipment.

Certain operations, because of their nature and location, may require an enclosure to isolate the operation from other personnel in the area.

Machine Guards

Rotating power brushes and buffs should be used only on machines that are equipped with safety guards, and these guards must be kept in place at all times.

Starting

Jog the machine before starting, to determine if it is ready to use and that the product is fastened securely. Run at operating speed with safety guards in place for at least one minute before applying work. Do not stand in front of or in line with the product during this time.

Pressure

Avoid excessive pressure against the work. This reduces the efficiency and could cause premature failure during operation.

Safety Standards

Comply with the Safety Standards of the Industrial Division of the American Brush Manufacturers Association and the American National Standards Institute B165.1-1985, Safety Requirements for the Design, Care and Use of Power Driven Brushing Tools, and B165.2-1982, Safety Requirements for the Design, Care and Use of Power Driven Brushing Tools Constructed with Wood, Plastic, or Composition Hubs and Cores.

APPLICATIONS OF BRUSH AND BUFF TOOLS

The best and most practical way to learn and understand the application of brush and buff tools is to examine the following series of photographs of existing in-plant operations (Figures 5.1–5.86). Review the captions under the figures for technical data and cost information for each of the operations.

OPERATION CHECKLIST

Remembering that wire brush tools are effective with the tips only and that abrasive monofilament tools utilize tips as well as sides, some helpful suggestions for users of brush tools which have been developed through state-of-the-art practices are in Table 5.1 on p.p. 134–135.

Figure 5.1   Deburring and edge blending steel helical gears. Brush used: two only 14 inch diameter Tufmatic knot type with 0.014 inch diameter high-tensile steel wire rotating at 1750 rpm. Brush time cycle: 10 seconds total per gear side. Brush cost: approximately $0.012 per gear side.

Figure 5.2 Cast steel helical gears showing before (top) and after (bottom) burr removal.

Figure 5.4 Recessed internal helical ring gear of a steel pinion housing part being deburred and having sharp edges removed on an automatic machine with one of two brushing heads shown. Burrs are turned into the gear helix by flat surface machining. Brush used: two only 6 inch diameter cup-knot type with 0.014 inch diameter, high-tensile steel wire rotating at 3450 rpm. Brush time cycle: 8 seconds each brush simultaneously. Brush cost: approximately $0.009 per gear side.

Figure 5.3   Steel pinion helical gears showing after (left) and before (right) burr removal.

Figure 5.5   Steel pinion housing part with a recessed internal helical ring gear showing the top profile deburred and sharp edges removed.

Figure 5.6 Steel pinion housing part shown on an automatic machine positioned to be brushed, deburred, and have its spline end profiles edge blended. Part rotation by the machine is 30 revolutions per minute. Brush used: two only side-by-side 14 inch diameter Tufmatic knot style with 0.014 inch diameter, high-tensile steel wire rotating at 1750 rpm. Brush time cycle: 6 seconds total. Brush cost: approximately $0.008 per part.

Figure 5.7   Steel pinion housing part with a turned-in machining burr on the end of a broached spline outer face. The burrs must be removed and the spline end profiles must be edge blended.

Figure 5.9   Internal helical ring gear of a steel housing part showing the burrs removed and sharp edges and corners rounded and blended to specifications.

Figure 5.8   Internal helical ring gear of a steel housing part showing the small, practically fractured, turned-in machining burr, with sharp edges and corners on the tooth profile areas.

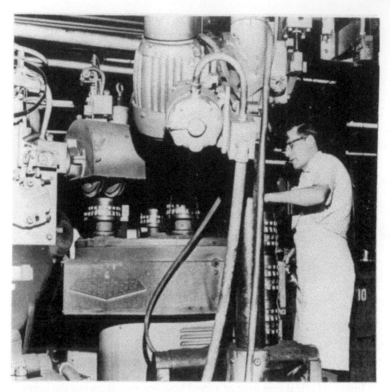

Figure 5.10   An overview of a semiautomatic, six-spindle, manually tended machine with two brushing heads. The internal helical ring gear on a steel housing part is being deburred with the sharp edges and corners rounded and blended. Two identical heads are used to accommodate the high production rates required.

Figure 5.11   A close-up view of the two brush heads used to deburr and round the sharp edges and corners of the internal helical ring gear on a steel housing part. The four brushes used on each of the two heads are spaced apart and paired together to uniformly cover the gear profile area. Brush used: four only 14 inch diameter Tufmatic knot type with 0.014 inch diameter, high-tensile steel wire rotating at 1750 rpm. Brush time cycle: 4 seconds contact with each head simultaneously. Brush cost: approximately $0.008 per part.

Figure 5.12   Steel pinion housing part on a semiautomatic machine with two 6-inch diameter wire wheel brush heads deburring the underneath side of the recessed internal helical ring gear. Being a confined and geometrically restricted area dictates using small-diameter, narrow-width wire wheels that are able to physically reach and deburr the hard-to-get-at burrs. Brush used: one 6 inch diameter knot-type wheel with 0.014 inch diameter, high-tensile steel wire rotating at 3450 rpm on each head. Rotation of each wheel is opposite to ensure burr removal from both tooth profiles. Brush time cycle: 10 seconds each brush simultaneously. Brush cost: approximately $0.011 per steel pinion housing part.

Figure 5.14   A cast metal hydraulic flow plate housing part being unloaded from a dual-head semiautomatic deburring machine. The housing parts are loaded and unloaded at the operator station with the machining burrs removed at the two brushing stations (behind protective shield); one brush is rotating in the forward direction, the other in a reverse direction. Brush used: two only 6 inch diameter cup, knot style, with 0.014 inch diameter, high-tensile steel wire rotating at 3450 rpm. Brush time cycle: 6 seconds each brush simultaneously. Brush cost: approximately $0.007 each housing part.

Figure 5.13 Two small steel helical pinion gears highlighting the condition of burr before wire brushing (left) and the deburred surface afterward (right). The spur engagement gear (pointed teeth) restricts the wheel brush from attacking the burrs at the desired position. A compromise by setting the wheel brush at an angle is sufficient to perform an adequate job.

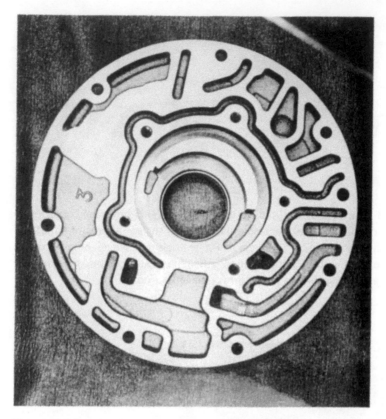

Figure 5.15   The brush-finished surface of a cast metal hydraulic flow plate housing part, showing the machining burrs removed and retaining the edges sharp for smooth flow characteristics of the finished assembly.

Figure 5.16 An overview of a six station automatic spur gear deburring machine with automatic load and unload stations and two wide-face wire brush deburring stations as shown. Three 12 inch diameter heavy-duty crimped-type wheel brushes are assembled side by side on each of the two deburring stations, providing wide coverage of the gear profile and minimizing the time cycle for the deburring and edge blending operation. Brushes used: two heads of three only, each crimped style with 0.014 inch diameter, high-tensile steel wire rotating at 1750 rpm, each head powered by a 5 horsepower motor (1 HP per inch of brush face). Brush time cycle: 4 seconds each brush head, simultaneously. Brush cost: approximately $0.014 per part.

Figure 5.17   A close-up view showing one of the two wire-brush deburring stations in operation. The 6 inch wide brush covers most of the width of the spur gear, providing maximum work per unit of time during the deburring cycle.

Figure 5.18   Dual integral steel helical pinion gears, illustrating conditions before deburring (left) and after deburring (right). The burr removal on the top profile surface of the lower gear, being angled, presents a more difficult removal task because of the restricted space limitations. A narrow-width single-wire brush wheel is positioned at an angle to attack the burrs and remove them.

Figure 5.19 A six-station semiautomatic indexing machine with three of the four wire deburring heads shown. The two heads positioned to deburr the top of the pinion gear part require two brushes side by side on each head. However, the top surface of the lower gear utilizes a single-wire brush wheel on each of the two heads, one operating in the forward rotational direction, the other in reverse direction. Brushes used: six only 14 inch diameter twist-knot type with 0.014 inch diameter, high-tensile steel wire rotating at 1750 rpm. Brush time cycle: 4 seconds total at each station operating simultaneously. Brush cost: approximately $0.018 per gear part.

Figure 5.20 The operator load/unload station showing the dual integral steel helical pinion gears being unloaded and loaded by the operator. Other sizes of gears are deburred on the same machine utilizing the wire brushes, but repositioned for the new deburring specifications.

Figure 5.21 A fully automatic six-station indexing machine with two abrasive monofilament wheel brush heads deburring and edge blending spur gear tooth profile surfaces (burr side only). Burrs of moderate to light adherence are better handled by the abrasive grinding brushes because the resultant surface finish of the edge blend and economies of the operation. Brushes used: two only 12 inch diameter crimped-type abrasive monofilament wheels, one on each brush head, rotational speed 1750 rpm. Brush time cycle: 6 seconds total at each station operating simultaneously. Brush cost: approximately $0.007 per gear part.

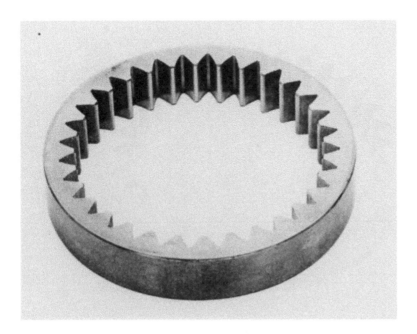

Figure 5.22 Steel internal ring gear, flat ground on both sides with light-abrasive-generated burrs rolled into gear teeth. This close-up view is before being deburred with a flexible abrasive finishing tool.

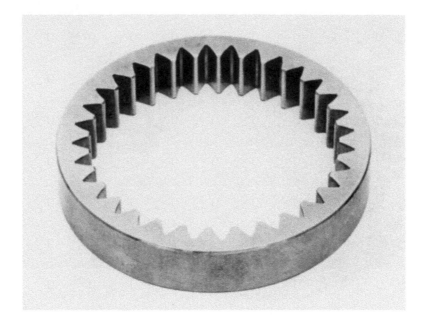

Figure 5.23   Steel internal ring gear, flat ground on both sides, illustrates the finished condition with burrs removed and tooth profile edges radiused with a flexible abrasive finishing tool.

Figure 5.24   A semiautomatic twin brush head indexing machine fixtured for deburring and edge radiusing automotive steel internal ring gears. The two wheel brush heads with flexible abrasive wheels rotating in opposite directions provide complete and uniform gear coverage. Both sides of each gear are uniformly deburred and radiused by utilizing the machine operator to load, turn over, and unload each gear. Brush used: one 12 inch diameter flexible abrasive finishing tool on each head utilizing a 320 silicon carbide grit nylon filament at 1750 rpm. Brush time cycle: 6 seconds total at each station operating simultaneously. Brush cost: approximately $0.008 per gear part.

Figure 5.25 A typical steel helical gear with moderate to heavy burrs (left) and the same gear with the burrs removed right. One of the desirable features of wheel brushes is the flexibility of the wire or abrasive fill materials. Not only is the flat tooth profile deburred, but also the 45 degree angle between the flat and the periphery of each tooth.

Figure 5.27   Removing weld spatter and rust (surface oxidation) from a steel fabrication with a 6 inch diameter, crimped 0.020 inch diameter, high-tensile steel wire brush rotated by a 4500 rpm portable power tool.

Figure 5.26   Removing weld spatter and slag from a heavy steel fabrication utilizing a 5 inch diameter with heavy-duty twist-knot 0.020 inch diameter, high-tensile steel wire brush rotating at 4500 rpm powered by a portable power tool.

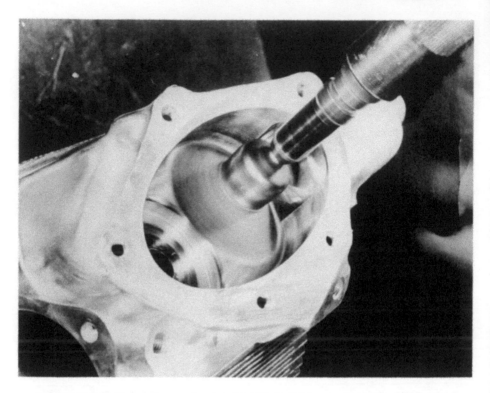

Figure 5.28   Deburring and edge blending the internal machined surface edges of a small aluminum engine casting. The 2 inch outside diameter, polyurethane encapsulated, 0.010 inch diameter, high-tensile steel wire brush operates at 20,000 rpm.

Figure 5.29   Centerless brush deburring, edge contouring, and surface conditioning of cylindrically shaped parts with holes, slots, and grooves, where a fine grinding burr is turned into the cavity by the preceding external outer diameter grinding machine operation. The large-diameter, wide-face, flexible abrasive wheel utilizes a 0.022 inch diameter, silicon carbide 320 grit filament with a compatible grinding coolant. Cost per part deburred is negligible as the operation is harmonized with a series of in-line grinders operating on a continuing throughput basis.

Figure 5.30   Illustrated is a semiautomatic continuous rotary table with three wire brush heads, two of which are positioned to pre-deburr the edges of broached slots on a steam turbine wheel, while the brush head in the rear pre-deburrs the peripheral diameter. The 14 inch diameter wheel brushes are rotated at 1750 rpm, utilizing 0.014 inch diameter, twist-knot, high-tensile steel wire.

Figure 5.31   End spline deburring and cleaning operation on an automatic straight line indexing machine. The steel drive shaft is in position to be clamped and rotated, and wire brushes by the radial wheel brush are directly above the spline.

Figure 5.32   The spline end of the steel drive shaft is clamped and rotating underneath the wire brush, which has been brought in contact with the spline end for the deburring and cleaning operation. A finished part is shown in the foreground. Brush used: one only 12 inch diameter narrow face wheel with crimped 0.010 inch high-tensile steel wire. Single direction of brush rotation at 1750 rpm. Brush time cycle: 6 seconds total. Brush cost: approximately $0.002 each part.

Figure 5.33 Steel sliding clutch plate with external teeth having medium-sized broached burrs (right). Completely deburred clutch plate (left) with burrs removed and sharp edges radiused.

Figure 5.34   Close-up view showing one of two brushing heads on a semi-automatic machine used to deburr steel sliding clutch plates. Two brushes are used on each head and positioned side by side as shown. Brushes used: four only 14 inch diameter twisted-knot type 0.014 inch diameter, high-tensile steel wire rotating at 1750 rpm. Brush time cycle: 6 seconds per part, brush heads operating simultaneously. Brush cost: approximately $0.016 each.

Figure 5.35 Continuous through-feed automatic centerless machine wire brush deburring automotive engine pistons. The centerless brush deburring is the final operation after a series of abrasive grinding machines grind the pistons to size. The brush used is 20 inch diameter by 6 inch wide with 0.008 inch diameter, high-tensile carbon steel wire operating at a rotational speed of 1200 rpm. Grinding coolant is applied to the brush and piston parts. Being a very fine deburring operation, brush life is measured in months with brush cost of deburring infinitesimal.

Figure 5.36 Two outside diameter steel drive spline power transmission parts, the one on the right showing the turned-in machining burrs on the spline face and inner edge. The part on the left illustrates the finished deburred spline face and inner edge.

Figure 5.37   A six-spindle semiautomatic indexing machine with three wire brushing heads (behind the enclosure). The machine operator loads, turns parts over, and unloads each part to obtain complete deburring of both spline faces and the outside diameter sharp edges. The three brushing heads are shown in the next three pictures. Brushes used: six only 14 inch diameter twist-knot construction wheels with 0.014 inch diameter, high-tensile steel wire. Brush rotation is single direction at 1750 rpm. Brush time cycle: 6 seconds total with all three heads operating simultaneously. Brush cost: approximately $0.024 each part.

Figure 5.38   The first brush head with two 14 inch diameter twist-knot wire brush wheels, spaced apart 2 inches, doing the spline face.

Figure 5.40   The third brush head with two 14 inch diameter twist-knot wire brush wheels, spaced apart 3 inches, removing the sharp spline edges on the outside diameter of the drive spline.

Figure 5.39   The second brush head with two 14 inch diameter twist-knot wire brush wheels, spaced apart 2 inches, deburring the spline face and inner edge.

Figure 5.41   Seal retained parts (steel) having the milled burrs removed from the outside diameter of the lower skirt area. One of the two brushing heads is shown in the rear, having two 14 inch diameter wire wheels side by side and slightly angled to cover approximately 2 inches of the burr area on the lower skirt. The other brush head is set up in a similar fashion except the brush rotation is opposite to ensure that the burrs on all edges of the skirt are removed.

Figure 5.42 A special automatic machine with one of the two 4 inch small diameter brush heads shown mounted on an air tool. The inner snap ring groove is visible on the inner/outer spline housing part, which has just been deburred by the brush wheel. During operation, the part is indexed underneath the brush. The brush automatically drops inside of the part and removes the machining burrs on the grooves. Brush used: two only 14 inch diameter crimped 0.014 inch high-tensile steel wire wheels, operating at 12,000 rpm. Brush time cycle: 8 seconds total with both brushes operating simultaneously in opposite directions. Brush cost: approximately $0.007 each part.

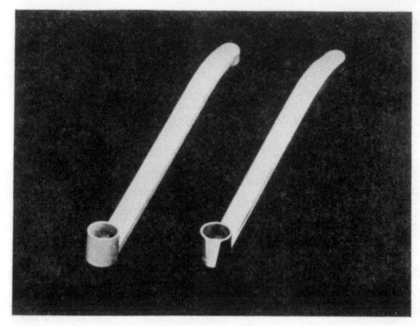

Figure 5.43   Buffing the exterior surfaces of single-lever tubular brass faucet spouts to a bright mirror finish. Brass faucet spout after bright dipping and before buffing on the left, with the completely buffed spout on the right.

Figure 5.44   The spouts are fixtured vertically on chain-driven platens that move past thirteen buffing heads on an automatic in-line buffing machine. The part areas are completely covered with the wide-face 22 inch diameter and 16 inch diameter wide-face bias buffs used with pressurized liquid "cut" and "color" compounds on the proper buff heads. The buffing time cycle is approximately 12 seconds per part with all heads operating simultaneously.

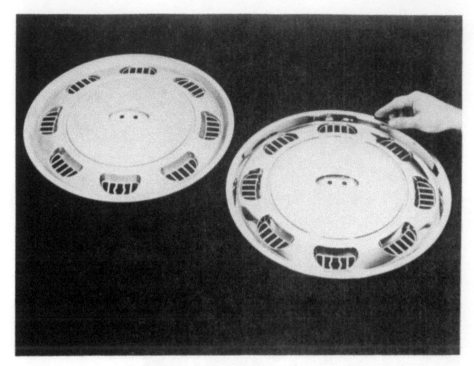

Figure 5.45   Cut and color buffing the top surfaces of stainless steel automotive wheel covers. The stainless steel cover in the stamped and not buffed condition is shown on the left, with the finished buffed, mirror-like cover on the right.

Figure 5.46 The wheel covers are manually loaded on horizontal fixtures that orientate and hold the parts during their rotation underneath six 24 inch diameter by 7 foot wide bias ventilated mush buffing wheel heads. Each buff section is spaced on the buff head spindle to provide flexibility for covering all top surfaces of the wheel covers. Separate pressurized liquid ''cut'' and ''color'' compounds are used—the first three heads cut buffing and the last three color buffing, with the direction of buff head rotation being opposite for each of the two series of heads. The buffing time cycle is approximately 4 seconds per part with all heads operating simultaneously.

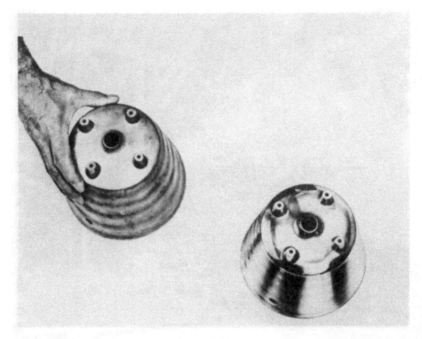

Figure 5.47   Appliance industry zinc die-cast blender bases being shown with the as-cast unbuffed part on the left and the cut and color buffed part on the lower right.

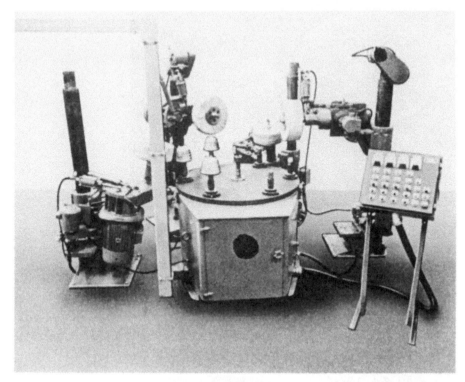

Figure 5.48   The parts are manually loaded and unloaded on a six-station rotary index machine with four 18 inch diameter by 6 inch wide bias buff heads contouring around the outer surfaces of the bases. An automatic liquid compound system is used, with the first two heads "cut" buffing and the last two "color" buffing at approximately 4000 sfpm. The buffing time cycle is approximately 10 seconds per part with all heads operating simultaneously.

Figure 5.49   Two automotive stainless steel wire wheel trim ring stampings with a multitude of sharp holes on the inner surfaces. The stamping on the left shows the condition of the part before buffing and the view in the lower right illustrates the finished trim ring "cut" and "color" buffed with a bright finish and all sharp edges removed.

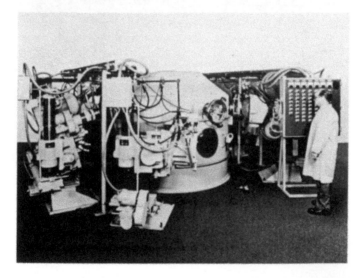

Figure 5.50   A manually loaded eight-station indexing machine with seven buffing heads: three utilizing ventilated sisal buffs for "cut" and four having ventilated bias buffs for "color" with wheel diameters suitable for 13 to 15 inch trim ring diameters. A dual low-pressure automatic compound system is used to continuously supply all buffing heads with adequate compound. The buffing time cycle is approximately 9 seconds per part with all heads operating simultaneously.

Figure 5.51 A close-up view of one ventilated sisal buff head cut buffing the trim ring stamping.

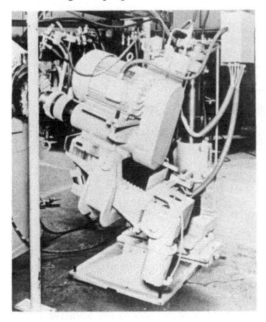

Figure 5.52 A close-up view of one ventilated bias buff head "color" buffing the trim ring stamping.

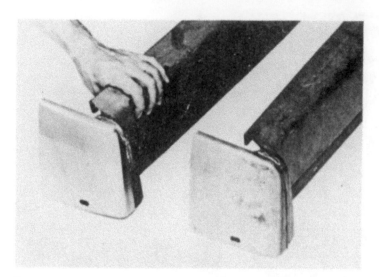

Figure 5.53   End finishing of automobile bumpers with a conveyorized polishing and buffing machine. A finished, polished and buffed bumper end is shown at the left, while the unfinished bumper end is on the right.

Figure 5.54   Four polishing and buffing heads are used on each side of the line, including two heads with 16 inch diameter abrasive flap wheels, two heads with 18 inch diameter sisal buffs, and four heads of 18 inch diameter cotton bias buffs. An automatic liquid compound system applies the buffing compounds, while a separate system feeds compounds to the abrasive flap wheels. The polishing and buffing time cycle is approximately 14 seconds per bumper with all heads operating simultaneously.

Figure 5.55 Buffing automotive steel and/or aluminum bumper surfaces on a manually loaded, gantry type, conveyorized machine.

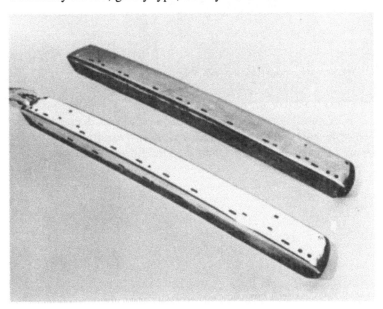

Figure 5.56 Two stamped aluminum bumpers are shown, with the rear one unpolished and the other in the front final finished.

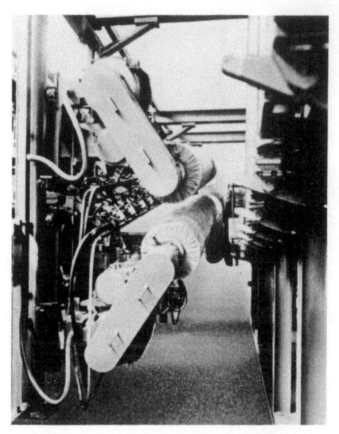

Figure 5.57   Six 12-foot-long mush buffing heads with 24 inch diameter bias buffs cut and color the bumpers to the desired surface finish requirements. Two cut buff heads on the left side of the machine start the finishing process.

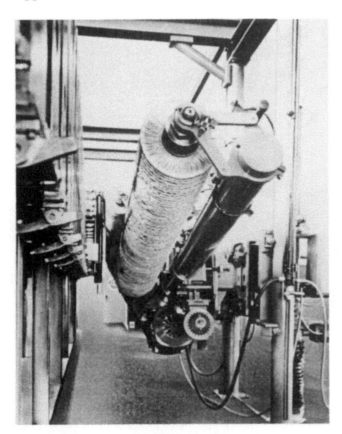

Figure 5.58   The buffing is finished by one pair of cut buffs (angled) and one pair of color buffs (angled) on the right side. Low-pressure liquid cut and color compounds are applied to the buff heads automatically. The buffing time cycle is approximately 6 seconds per bumper surface with all heads operating simultaneously.

Figure 5.59 Pictured is a robot holding and manipulating a lightweight plastic magnetic disc storage system part (3 inch cube) in contact with a fixed position rotating flexible abrasive/nylon tool. The 12 inch diameter by 1 inch wide wheel performs a precision fine parting line removal and surface blending operation, with 240 abrasive grit aluminum oxide, rotating at 900 rpm. The robot end effector puts the part through a programmed path within the envelope of the cube.

Figure 5.60   Robot deburring and edge contouring using a 22 inch diameter, 12 inch wide, 80 grit silicon carbide flexible abrasive tool deburring and edge blending aluminum aerostructure parts mounted on a four-position work-holding fixture. The wheel brush tool design consists of 20 strips of 8½ inch long trim length material mounted around a center hub, to reach and blend out the milled mismatch lines on the bottoms of the 4 to 6 inch deep cut-outs. Utilization of the silicon carbide and nylon material sides to wipe across all the upper and top edges removes the milling burrs and provides the necessary edge blends to specified tolerances.

Figure 5.61   Close-up view of the 22 inch diameter by 12 inch wide flexible abrasive tool mounted on a robot, deburring and edge contouring an aluminum crescent-shaped aerostructure part. With a rotational speed of approximately 450 rpm, combined with deep abrasive filament penetration of about 4 inches, the continuous wiping action of the abrasives across the sharp burrs grinds away the burrs and produces a smooth consistent edge contour to the part specifications. The programmed robot generates the tool path and follows the outline of the part contours, giving repeatable consistent coverage.

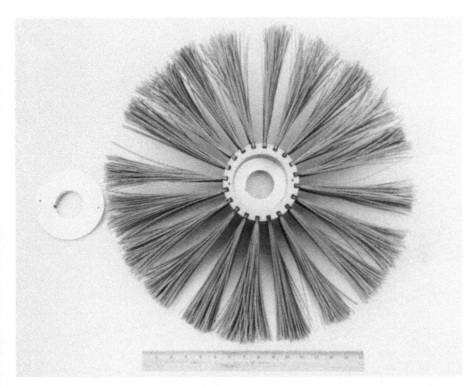

Figure 5.62 Pictured is a new 22 inch diameter rectangular abrasive and nylon finishing tool, which is replacing the conventional abrasive wheels used for deburring, radiusing, edge contouring, and surface conditioning work. Designed with an aluminum center hub having 20 slots, replaceable strips of rectangular abrasive/nylon filaments with various grain types and sizes are inserted and contained by the end flanges. Primarily, usage is on large robots and machine tools having the capacity to utilize abrasive tool widths up to 36 inches. Large radial wheel sections, $\frac{1}{2}$ inch wide, are also used in wide-face cylinder form with spacers for light or medium densities and/or placed together for heavy density applications.

Figure 5.63   Pictured is a 14 inch outside diameter by 5 inch wide ''long string'' (long length filament) strip finishing tool utilizing the new rectangular abrasive silicon carbide or aluminum oxide monofilaments. Application experiences have shown this material to be superior to the conventional round monofilaments by providing full line contact with the work, not point contact. With a total weight of 18 ounces and operating at 2200 sfpm, the tool is ideally suited for small-sized robots and/or automated machinery.

Figure 5.64   Flexible abrasive finishing tools provided a productive alternative to deflashing and edge contouring composite materials. Pictured is a "before" (top) and "after" (bottom) view of an uncoated automotive fiberglass composite door panel showing the excess flash. After deflashing and edge contouring with abrasive finishing tools, the finished part edge is illustrated with a uniform, controlled, edge contoured finish. The impact of the abrasive filaments and the grinding action of the abrasive minerals combine to achieve a uniform edge finish on a productive basis.

Figure 5.65   The illustration shown here is similar to Figure 5.64 except that the automotive composite door panel is presented to the robotic tool in a coated form. The straight white line on the bottom part shows the uniform flash removal and edge contouring results. Abrasive tool diameters of 14 to 18 inches operating at surface speeds of 2000 to 3000 sfpm provide the most economical performance and lowest end-of-service costs.

Figure 5.66   Illustrated is a new 12 inch diameter disc wheel using 36 strips of rectangular abrasive nylon filaments inserted into an aluminum disc to make a flat sheet or flat surface deburring, edge contouring, and finishing tool. Disc diameters can range from 8 to 48 inches in diameter and provide a circular rotational motion to achieve a 360 degree contact wih the areas or surfaces to be deburred or surface conditioned. Rotational motion can be in either direction or orbital. New applications include the deburring of large outer case parts with scallops or cutouts, hydraulic valve bodies, machined transmission cases, flat sheets and plates, and double-disc ground parts.

Figure 5.67   This view shows the working face of the disc wheel with a 4 inch long by 2 inch high strip that makes up the wheel assembly. The strip lengths vary depending on the disc diameter and area to be covered. The strip height is designed to accommodate the stiffness of the filaments and space limitations of the piece parts to be finished. Strip sets can be provided in silicon carbide, aluminum oxide, and other special minerals in a wide range of grit particle sizes. Densities can be modified by varying the number of strips used in the disc. Operational speeds are in the area of 2000 sfpm, and provisions are made to use coolant internally fed through the drive spindle.

Figure 5.68   Honing the sharp edges of both sides of carbide tool inserts is usually performed with crimped, round, abrasive and nylon filaments as illustrated. The insert is positioned at least half the thickness into a rotating pocketed fixture with the filaments sweeping across the insert edges while touching the recess angle of the fixture. The purpose of honing is to generate a uniform 0.001 to 0.002 inch edge radius to prevent stress cracks occurring and to ensure better high-temperature coating adherence. The inserts are turned over in the fixture to hone the opposite side.

Figure 5.69   An overview of the continuous honing process showing two wide-face flexible abrasive wheel heads positioned over the conveyorized multiple spindles holding the carbide inserts. The closed-face construction wheel heads are made up of many narrow-width individual sections placed together to achieve the maximum abrasive/nylon filament contact with the carbide inserts edges. Peripheral speed for the round filaments ranges between 3700 and 4400 rpm. No coolant is used.

Figure 5.70   Two carbide insert tools are pictured with the edges uniformly honed. The top has an overlay of a single round, crimped, abrasive/nylon filament. Similarly, the bottom has a single rectangular, flat, abrasive/nylon filament shown. The comparison made here is based upon identical machine operations using flexible abrasive finishing tools having the same diameters, wheel speeds, insert rotational speed, and overall time cycle, with the significant difference being in the round versus rectangular abrasive/nylon filaments used. Summarizing the highlights, this focuses on the advantages of the new advanced rectangular filaments because the edges were uniformly honed twice as much, with one-quarter wheel pressure (infeed contact) and using one-half the wheel widths. With the experiences gained on the other deburring and edge contouring applications and the much-increased performance of the honing operations, wheel rotational speeds should be lowered to the range of 2500 to 3000 sfpm, thereby being more cost effective in terms of end-of-service.

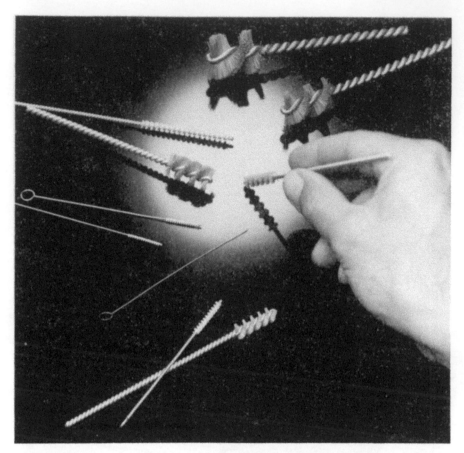

Figure 5.71 Illustrated is a group of precision microabrasive tools utilizing twisted-in wire construction with diameters from 0.030 to 1.015 inches. Product applications are for deburring, edge breaking and surface cleaning of internal drilled, milled, broached, and cut holes, slots, bores, cavities, threads, and hollows. The flexible abrasive materials used are a very fine alumina/silicate mineral and/or a fine 600 grit aluminum oxide encapsulated in flexible and resilient strands of special nylon. Polycrystalline diamond material is also available and suitable for special super alloys and advanced composites. Rotational speeds from 1000 to 2500 rpm are usual and are compatible with NC and CNC machine speed ranges.

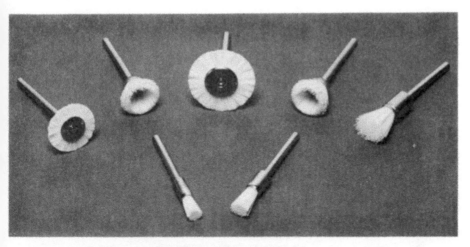

Figure 5.72   Miniature abrasive tools ranging in sizes from $\frac{1}{4}$ to $1\frac{1}{2}$ inches in diameter utilize the very fine alumina silicate abrasive/nylon material as used in the microabrasive tools. Three standard shapes include radial and cup wheels plus the end spot facing type. The most effective operational speeds are between 1000 and 2500 rpm for all metals, plastics, and composite materials. Applications include very light deburring, edge breaking, and surface conditioning in the computer, electronic, medical equipment, hydraulic, and aerospace component parts industries.

Figure 5.73 A close-up view of a microcomputer part with a 0.125 inch diameter microabrasive tool inserted through the center hole. The very fine 0.008 inch diameter silicate abrasive/nylon material is used to deburr the internal cross slot (middle) fine pitch threads of both ends and not change the 4 rms finish on the honed bore. The operation is performed on semiautomatic equipment, which rotates and powers the tool through the internal surfaces. Microabrasive tools similar to these are now being used in increasing volume on NC and CNC machines, incorporating the deburring and cleaning operation into the programmed machine cycle.

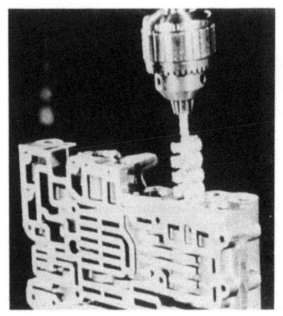

Figure 5.74   Pictured is the deburring and edge breaking of internal intersecting holes in a machined aluminum hydraulic valve body part. The parts are fixtured and processed semiautomatically on a drill press machine. The 1 inch diameter, aluminum oxide, flexible abrasive tool is operated between 1500 and 3000 rpm while powered in and out of the holes. The light drilling burrs are removed and a less than 0.001 inch edge break is generated at the sharp corners. The surface rms can be maintained depending on the abrasive mineral and size used.

Figure 5.75 Numerical controlled (NC) and computer numerical controlled (CNC) machines are being used in conjunction with flexible abrasive finishing tools to incorporate the deburring, edge contouring, and surface conditioning of piece parts after the conventional machining, drilling, and milling cycles. The abrasive tools are chucked in tool holders, placed on the tool changer, and programmed into the machine sequence and overall work cycle. Shown here is a microabrasive tool, 0.075 inch diameter, used to remove internal cross hole burrs on aluminum hydraulic valve body parts after the drill and ream cycle. The small burrs are removed, sharp edges contoured less than a thousandth of an inch and any residual sharp surfaces polished out. Rotational speeds are 2000 rpm.

Figure 5.76 Small hole and thread deburring, polishing, and cleaning are accomplished by using one of the many flexible abrasive helical finishing tools shown above. Various tool sizes range from $\frac{3}{8}$ inch in diameter to as large as $1\frac{1}{2}$ inch diameter with incremental sizes in between. The open-face flexible construction permits complete abrasive contact with the root and peaks of threads, in addition to deburring cross holes or slots in conventional holes.

Figure 5.77   Cylinder bore deburring, prior to honing, can be performed by the use of the large-diameter flexible abrasive finishing tool, designed and made with the new flat/rectangular abrasive nylon monofilaments. The 4½ inch diameter size shown is used for a 3½ inch diameter bore, thereby producing a ½ inch line contact of the flat filaments against the side walls of the cylinder bore. When encountering the cylinder fuel ports which need deburring and edge contouring, the ½ inch length of flat filaments drops into the port opening during rotation, and upon exiting draw files itself across the burrs to remove them and continues to generate the required edge radius or contour.

Figure 5.78   The new tool on the left is compared to the one on the right, which has been used to remove all burrs, chips, and other particulate matter on two-cycle engine blocks. Using the flat/rectangular abrasive nylon filaments with 120 grit aluminum oxide, at 600 rpm and 30 reciprocating strokes, a total of 120 engine blocks were finished. This compared very favorably to the old point contact method, which utilized plastic/vitrified tipped brushes of the same grit, slower rpm, 20 reciprocating strokes, and finished only 12 engine blocks per brush.

Figure 5.79   End or spot facing abrasive tools, with the new rectangular-shaped abrasive filaments, are used in conjunction with NC and CNC machines to deburr and finish larger internal and external surfaces. Abrasive tools similar to this range in diameter sizes from the miniature abrasive ¼ inch to the standard 1½ inch diameter with abrasive filament lengths from ¼ inch to 4 inches long. Tool rotational speeds are suitable for NC and CNC machines, with or without coolant.

Figure 5.80   End or spot facing type tools are used to deburr and polish blind hole surfaces that are inaccessible by other tools. The tool shown has a removable ½ inch wide plastic sleeve or bridle restricting the flare of the abrasive filaments to the tips only. The added stiffness of the filaments provides a very aggressive work factor to the parts. In some cases, the plastic bridle doubles as a "bumper" to guard against the metal part of the tool coming in contact with the part surface and causing scratches or marks.

Figure 5.81 Cup wheels with $\frac{1}{4}$ inch mounting shanks begin with the $1\frac{1}{2}$ inch diameter sizes and range up to 8 inches diameter with incremental sizes in between. The lengths of abrasive/nylon filaments (trim) vary from $\frac{1}{2}$ to 5 inches long and utilize both the round-shaped material for medium flexibility and cut and the rectangular for greater stiffness, more aggressive cutting. The larger cup wheels with silicon carbide abrasives perform very well on CNC machines that turn, mill, and bore large aluminum castings. Deburring and edge contouring feed rate averages in the area of $2\frac{1}{2}$ feet per minute using the available coolants.

Figure 5.82 A 3 inch diameter flared flexible abrasive cup wheel is shown attached to a tool holder. The flared shape permits cornering of the abrasive filaments to deburr and clean shallow angle corners and difficult geometric part shapes. In addition, when pressured against flat or moderate irregular shaped surfaces, excellent cutting is produced across a wide area; for example, the 3 inch diameter tool will cover a 3 inch wide path on the surface of the part or parts.

Figure 5.83 Radial wheel tools 2, 3, or 4 inches in diameter and $\frac{1}{8}$ to $\frac{1}{2}$ inch wide are used to deburr, edge contour, and refine external and internal part surfaces while on CNC or NC machine tools. Diameters, other than those listed, can be made to accommodate specific bore or hole sizes and/or restrictive outside part geometries. Rotational speeds should be in the 2500 to 3000 sfpm range for the most productive and economical results.

Figure 5.84 A five-axis computer numerical controlled (CNC) machine tool is illustrated utilizing a 2 inch diameter flexible abrasive finishing cup wheel in the tool holder. A cluster of small precision-machined, drilled, and tapped aluminum missile component parts are being deburred and radiused while still in the holding fixture and directly after the last machining operation. It makes good sense to do as many machining and finishing operations as possible on the machine. Expensive setup time is minimized and inefficient secondary operations are eliminated. In addition, the combining of finishing operations with machining cycles produces uniform, predictable, timely, reliable, and productive results.

Figure 5.85   Shown here is a computer numerical controlled (CNC) machine tool utilizing a 2 inch diameter, 80 grit silicon carbide/nylon abrasive finishing tool positioned in the retracted neutral position. The tool is used to deburr and radius the edges of a grouping of small precision-machined, drilled, and tapped aluminum missile component parts while still in the holding fixture and following the last machining operations.

Figure 5.86 Pictured is a close-up view of the flexible abrasive finishing cup wheel in contact with the machined aluminum parts. Being 2 inches in diameter, the tool covers more than one part during the traverse of the programmed cycle. The rotational speed of the tool is 2100 sfpm and the same coolant used for machining is used for the deburring cycle. The 80 grit silicon carbide/nylon filaments of the cup wheel maintains and slightly improved the machined surface finish in the 32 rms range.

Table 5.1   Trouble-Shooting for Brush Tool Operations

Observed result	Corrections suggested (wire)	Corrections suggested (abrasive monofilament)
Brush tool works too slowly	Increase surface speed by increasing outside diameter, rpm, or both. Decrease trim length. Increase filament density. Increase filament diameter.	Increase surface speed by increasing outside diameter. Increase abrasive mineral size. Decrease trim length. Increase filament density. Increase filament diameter.
Brush tool works too fast	Reduce surface speed by decreasing outside diameter, rpm, or both. Reduce filament diameter. Reduce filament density.	Reduce surface speed by decreasing outside diameter, rpm, or both. Reduce filament diameter. Reduce abrasive grain size. Reduce filament density.
Action of brush tool peens burr to adjacent surfaces	Decrease trim length. Increase filament density. If metal is too ductile, change to nonwire brushes such as abrasive monofilaments.	If burrs are not removed, change to larger grit size. Increase filament diameter. Increase trim size.
Finer or smoother finish required	Decrease trim length. Increase filament density. Decrease filament diameter. Use abrasive monofilament brushes.	Decrease filament trim length and increase fill density. Reduce filament diameter. Reduce grain size. Use aluminum oxide minerals instead of silicon carbide.
Finish too smooth and lustrous	Increase trim length. Reduce brush filament density. Increase filament diameter.	Increase filament diameter. Increase grain size. Use silicon carbide minerals instead of aluminum oxide.

Table 5.1　*Continued*

Observed result	Corrections suggested (wire)	Corrections suggested (abrasive monofilament)
Brushing action not sufficiently uniform	Increase trim length. Decrease filament density. Use mechanical fixturing or machine operation to avoid irregular off-hand manipulation of brush tool or part.	Use finer abrasive grit sizes. Increase filament densities. Use mechanical fixturing or machine operation to avoid irregular off-hand manipulation of brush tool or part.

6

Emerging and Future Trends

We have reviewed the many variables and parameters surrounding the state of the art of mechanical deburring, edge contouring, and surface finishing as related to power-driven brush tools and buffs from the mid 1940s to the present time.

Much has transpired over the years in the design, engineering, manufacturing and applications of products to serve the needs of industry. With a new materials revolution now under way, the accelerated pace of matching advanced materials to precision industrial application technologies will be the foremost challenge of the future, specifically between now and the turn of the century.

Modern manufacturing systems and "factories of the future" can no longer use standard commodity brushes and buffs on sophisticated machine tools to perform critical, uniform-quality, repetitive, precision material removal and finishing of expensive, microtoleranced, super-alloy and advanced composite piece parts.

Applied science and engineering technologies are being developed and are needed to understand and apply the many emerging abrasive minerals and materials that will be used in the coming decade. The term brush or buff, as we know it today, will be transposed into "precision flexible abrasive finishing tools"—and rightly so, because the fill material in the tools is the "bridge" between the piece parts that require consistent quality deburring, edge contouring, and surface finishing and the advanced machine tools, integrated manufacturing cells, robotics, and automated processes and systems. The new advanced fill materials in precision flexible abrasive finishing tools are the key issue that will make the difference in improving productivity, elevating quality, and reducing end-of-service costs.

An excellent example of recent technology developments in abrasive filament materials is best illustrated by reviewing the following case history of an important industrial application.

CASE HISTORY

About 20 years ago an engineered plastic material with abrasive mineral homogeneously dispersed throughout the round cross-sectional area was introduced into the industrial scene. Disregarding the various trade names, the material is commonly referred to as an abrasive nylon monofilament.

Many physical characteristics were similar to the wire being used in that the round diameter and length of the monofilament related to stiffness; the hardness was built into the abrasive mineral for cutting ability; and there was a degree of flexibility without exceeding the modulus of elasticity, also related to filament life and subsequently product life.

The single and most important unique and outstanding difference was and is the ability to cut or abrade, not only with the tips of the filament, *but with the sides also.*

Because the abrasive nylon monofilament was developed and introduced as a replacement for wire filaments, synthetics, and some natural fibers, it remained in the shadows of mediocrity with slow growth. Reference was made to it as an expensive, hard to work with,

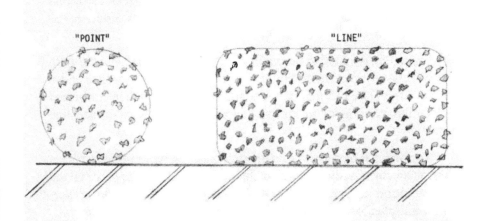

Figure 6.1   Surface contact: Round versus rectangular abrasive/nylon mono-filament.

and difficult material to cut or trim. Very few people researched, examined, tested, or understood the concept of maximizing the abrasive in the filament for deburring, edge contouring, and surface finishing.

The basic premise underlying the principal of developing the rectangular/flat shape flexible abrasive nylon monofilament is maximizing the "abrasive-on-the-surface" of the work piece, through *line* contact versus *point* contact (Figure 6.1).

When the new material became commercially available approximately 2 years ago it was designed, developed, engineered, and manufactured into a select group of flexible abrasive finishing tools without much in-house development and laboratory test work. Instead, after some basic theoretical studies were made, the product was put directly into existing production operations and tested by process engineers against the standard round/crimped abrasive filament products available at the time.

Immediately, the first production test results indicated a minimum of two to three times more cut, more uniform quality results, and a filament life of twice that of the standard material. This then confirmed the original premise of maximizing the abrasive-on-the-surface line contact.

The original premise of maximizing the abrasive-on-the-surface is based on an early study made of the theoretical differences between the standard round/crimped abrasive nylon filaments and the newly developed rectangular/flat abrasive nylon filament form.

Figure 6.2 illustrates the optical differences between both types of filaments against a backdrop of cutting tool carbide inserts.

Figure 6.3 presents the data on physical differences of the two filament types, which were used to determine the maximization of abrasive-on-the-surface and the advantages of line versus point contact using the sides of the materials. The ratio of abrasive-on-the-surface, therefore, equates to a theoretical 17.80 to 1.00 in favor of the rectangular/flat material.

One of the first laboratory tests performed to confirm and support the theoretical findings was done on the square-shaped, hardened, carbide cutting tool inserts illustrated in Figure 6.2.

During the manufacture of carbide cutting tools, one of the final finishing operations is to "hone" or grind a range from 0.002 to 0.005 inches radius on all sharp edges to eliminate stress risers, which can cause cutting tool fractures. In addition, radiused edges are necessary when hard coating the cutting tool inserts, because the coatings used will adhere better physically to radiused edges than to sharp edges. The radii requirements vary, depending upon the size, shape, hardness, and application of the inserts. In the case of the square inserts shown in Figure 6.2, the requirement was for a 0.002 inch radius on all sharp edges.

The purpose of the laboratory test was to hone or grind the edges

Figure 6.2 Optical differences between round/crimped abrasive nylon filament (top) and rectangular/flat abrasive nylon filament (bottom).

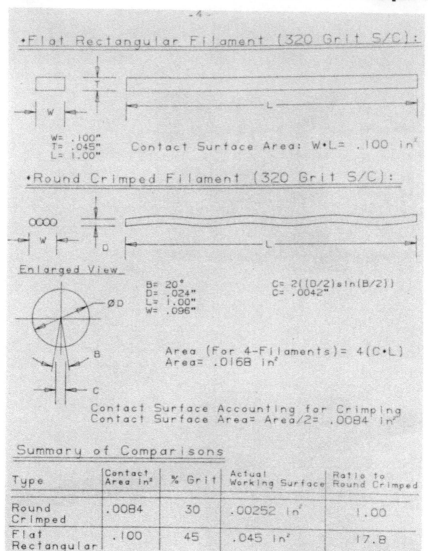

- 4 -

Flat Rectangular Filament (320 Grit S/C):

W= .100"
T= .045"   Contact Surface Area: W·L= .100 in²
L= 1.00"

Round Crimped Filament (320 Grit S/C):

Enlarged View

B= 20°   C= 2((D/2)sin(B/2))
D= .024"   C= .0042"
L= 1.00"
W= .096"

Area (For 4-Filaments)= 4(C·L)
Area= .0168 in²

Contact Surface Accounting for Crimping
Contact Surface Area= Area/2= .0084 in²

Summary of Comparisons

Type	Contact Area in²	% Grit	Actual Working Surface	Ratio to Round Crimped
Round Crimped	.0084	30	.00252 in²	1.00
Flat Rectangular	.100	45	.045 in²	17.8

Figure 6.3   Physical differences between rectangular/flat and round/crimped abrasive nylon filaments with actual working surface comparisons.

of six inserts for a 30-second time cycle each: three inserts with a round/crimped abrasive nylon filament tool and three inserts using the rectangular/flat abrasive nylon filament tool. The six honed insert radii were optically measured and the results were charted to determine the degree of work performed by the two different filament tools.

The test setup and operating conditions are best described pictorially in Figure 6.4. The test parameters were developed to simulate, as closely as possible, the actual production conditions existing in the field, based on using the data pertinent to the standard tools using the round/crimped materials.

Summarizing the laboratory test results, Figure 6.5 indicates a 2 to 1 ratio of edge honing for the rectangular/flat material. This is in spite of the fact that a filament penetration ratio of 1 to 4 *and* a points-per-square-inch ratio of 1 to 9.4 both favored the round/crimped material, as shown in Figure 6.4.

Here again, the initial laboratory test results on carbide cutting tool insert honing confirmed after the fact the first *production* test results on other application in the aerostructures/aircraft industry with the rectangular material providing 2 to 1 and 3 to 1 cutting ratios over the other materials. The test results also provided additional support for the abrasive-on-the-surface or line contact premise now supported by the theoretical analysis of a 17.8 to 1 ratio.

There still remained the question of proving product life on the carbide honing application, which in the long run translates to end-of-service costs. Therefore it was necessary to take the new abrasive tools into the field and place them on production equipment to ascertain actual life cycles. At the time, 80 hours of life was considered acceptable.

Contacts were made with several leading carbide cutting tool insert manufacturers and the data presented. Review by the manufacturing and process engineers led to on-line production testing over a period of several months. A summary of the customer's evaluation indicated that in general terms, over a period of months of operation, the following results are being obtained:

Tool life achieved is raised to a 3 to 1 or 4 to 1 ratio.

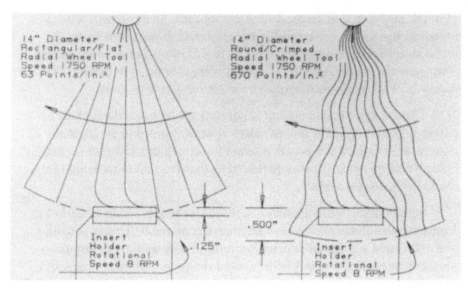

Figure 6.4 Lab test setup and operating conditions.

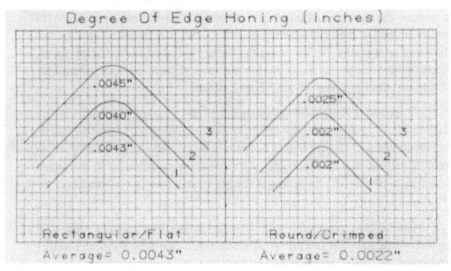

Figure 6.5 The measured honed radii on the six inserts used in the laboratory tests, each with a 30-second cycle time, produced an average result of 2 to 1 in favor of the rectangular/flat material.

Productivity or work per unit of time is doubled.

Product quality of honed parts is more uniform.

Less machine down time for abrasive tool changes.

More predictable and closely held tolerances because wear pattern of the abrasive is constant.

Rotational speed of the tools can be lowered without sacrificing performance, providing smoother wiping action of the filaments.

Less dry swarf generated—the abrasive minerals are held more securely in the larger cross-sectional area of the nylon component.

APPLICATIONS ON NC AND CNC MACHINE TOOLS

Another recent development of technology is happening in the machine tool industry. Flexible abrasive finishing tools are finding their way onto numerical controlled (NC) and computer numerical controlled (CNC) machine tools to deburr, edge contour, and surface finish piece parts after the turning, drilling, milling, and grinding cycles. This eliminates the secondary mechanical finishing operations and takes advantage of the single machine or cell doing all the programmed control, material removal from start to finish. Combining operations on one machine, with one set of tooling or fixturing, produces a more uniform, predictable, reliable, and productive result than previously obtained.

Application-wise, flexible abrasive finishing tools can remove burrs and stress risers and can perform radiusing, contouring, and edge blending. They can clean threads, slots, bores, and cross holes. Flexible abrasives can clean interior or exterior surfaces in geometric or irregular shapes. They can also be used for general surface conditioning and finishing.

Like most good tools, there are limitations to what can be accomplished with flexible abrasives. If burrs are heavy and inconsistent in size, the finishing operation will not produce complete removal and uniform results within an economical time cycle.

If there is pushed or extruded metal resulting from a worn drill bit breaking through at the intersection of two or more internal holes, the dwell time of the flexible abrasive tools may be excessive in order to grind the metal away. This situation may allow heat to build up when fine-sized abrasive particles are used, causing the nylon filament to soften. Selecting a medium or coarse abrasive particle may be helpful, but could cause detrimental effects if the surface finishes are in the precision range of 2 to 4 rms.

There are some characteristics to consider when radiusing, contouring edges or blending corners. On a sharp 90-degree corner, the initial contact of a flexible abrasive finishing tool will generate a ''true'' radius up to about 0.005 inches. Beyond that point, however, penetration of the abrasive filament decreases and a lateral wiping action on the top surface becomes greater, thereby producing a flattened edge contour in the range of 0.008 depth and 0.012 inches across the top. Longer dwell time and increased pressure of the tool further accentuate this effect.

Favoring a medium or coarse grain size will aid in holding closer to the true radius, but in some cases leaves a coarser finish.

Coolant and/or a lubricant is desirable and necessary to reduce heat, carry away the swarf, and provide more uniform part-to-part finish.

How are the deburring and finishing cycles integrated into NC and CNC machine operations? Before we answer this question, we need to understand the reasons why this automation is important. A consistent level of quality on a repeatable basis is better achieved on a machine-controlled operation than by manual means. Too many times very expensive precision-machined work pieces are scrapped due to inconsistent manual deburring and finishing.

In-process work flow can be much improved by performing the deburring and finishing within the machining cycle. There is no need to remove the machined work piece for the secondary operation. Moreover, performing all operations on the machine will increase throughput, improve delivery times, and put scheduling on a more predictable basis. Further efficiencies come from reducing or eliminating the manual deburring operations and department. Not only does it save some or all of the labor cost, but also most of the ineffective deburring materials, hand tools, and overhead associated with the manual deburring area.

There are a number of ways to integrate deburring and finishing cycles into the NC and CNC machines. The best place to start is to examine every deburring operation done manually. Find one piece part and one area that lends itself to deburring with flexible abrasive tools on NC machines. Determine if the speeds, feeds, time cycles, coolants, tool wear sensing devices, toolholders, and so on are compatible with the deburring tools.

The tool engineer and/or manufacturing engineer should monitor the initial operations to ensure that the work piece is deburred to specifications and tolerances and that the tool and machine function properly. After the first tool is successful, add another one or two to the cycle, and so on.

MODIFYING MACHINING APPLICATIONS

Another opportune time to consider flexible abrasive deburring tools is at the design phase of a new part. If possible, the tool engineer should work with the designer and tool supplier to provide as much application information as possible to help integrate deburring into the machining cycle. Minor design modifications may be necessary to accommodate this strategy. Before the new part is put on the machine for preproduction processing, the deburring tools and their programmed path and cycle time should be worked out just like the other machining and cutting tool programs.

The first preproduction run should be monitored closely. Then changes in the deburring tools, program, or cycle times can be made, if necessary. The more experience gained, the easier it is to bring new applications of deburring and finishing on-line.

There are a number of pitfalls that should be discussed at this point. Cutting tools performing operations prior to deburring must be monitored carefully for wear and changed at predetermined intervals. For example, a drill that wears sooner than anticipated and is not changed immediately may create a burr that is too heavy to be removed with flexible abrasive tools.

Internal cross-hole burrs pose a different problem for microabrasive hole deburring tools. If feed rates of cutting tools are increased to boost productivity, the intersection of a small hole into a larger one may not only leave a burr, but also distort the work piece. The abrasive tool will remove the burr but will not remove the distorted metal. Inspection gauging will reject the parts. Thus, downstream effect must be considered seriously when modifying a machining operation.

If abrasive filaments in the finishing tools cannot be used for whatever reason, a 0.005 inch diameter type 302/304 stainless steel wire or a nonabrasive loaded nylon in the finishing tools may be substituted. However, since both of these materials depend on the tips or ends of the filaments for impact and cutting action, they are less efficient than the abrasive-loaded filaments. Some new nonabrasive Aramid filaments have been introduced, and with their high-performance strength, stiffness, and heat resistance, they do well in special circumstances where metal burrs are of the "feather" type and on composites and advanced composite materials. These nonabrasive filament finishing tools can work with or without coolant, but an adequate supply of coolant is preferred. In some cases, a cool air blast can be applied at the contact point to reduce heat generated and removal of material particles.

Flexible abrasive finishing tools can be used in conjunction with other types of deburring tools on NC and CNC machines, and there are benefits derived from combining the best features of several finishing tools. For example, consider a heavy extruded burr on the entry or exit end of drilled or bored holes. The flexible abrasive tool will not remove the extruded metal part of the burr. However, a carbide cutter, countersink, or carbide burr will remove the extruded metal but will leave several small sliver burrs on the corners or surface juncture. By programming the machine to drill, chamfer with a carbide cutter, and then deburr with the abrasive tool, the finishing operation will be successfully achieved.

A similar operation is drilling, reaming, and using an abrasive finishing tool to remove residual metal slivers from the hole surfaces or to improve the finish. Precision honing of internal surfaces can be enhanced by a flexible abrasive finishing operation. The tool can be programmed to uniformly round the surface peaks without changing size or tolerances.

A combination of more than one flexible abrasive tool may be employed for deburring and surface finishing. Usually a coarse grain (80 mesh) tool removes the burr, and a medium grain (180 mesh) tool follows to blend away scratch marks and surface scratches.

The finishing tools being designed for today's machines encompass a wide range of new deburring materials, abrasives, shapes, and geometric forms. Superabrasives, such as polycrystalline diamond and cubic boron nitride, in nylon filament form are available and are now being used on a number of new applications.

SUMMARY

In reviewing the two major case histories of Mechanical Deburring and Surface Finishing just described, we need to summarize a number of pertinent points that are important to successful and productive uses on automated machinery, automated manufacturing systems, cells, robotics, and NC and CNC machines.

The flexible abrasive tools are selective in nature, as indicated throughout this publication. They work on specific areas to be deburred without having to mask off adjacent areas to protect from being scuffed or marred. The tools, being flexible, are more forgiving than hard grinding wheels or carbide burrs and less prone to part damage during machine malfunctions. They follow contours, round corners, and contour edges, as well as deburring and surface conditioning.

A wide selection of new abrasive minerals and grain sizes and shapes is available, including superabrasives, to engineer solutions to difficult application problems not solved in the past. It is the abrasive that does the work—the nylon is the carrier of the abrasive.

Peripheral speeds are normally slower than hard deburring tools and in the range of 2000 to 3000 surface feet per minute. The side wiping action of the abrasive filaments produces more cut and some heat generation. Slower speeds with the proper grain size abrasive and coolant or cool air application will minimize heat buildup, which is detrimental to the nylon in the filament.

Flexible abrasive finishing tools are not a panacea for all deburring problems. They are reliable, high-performance tools that produce consistent quality and cost-effective results when properly engineered for the specific applications intended.

Looking to the future, specifically the turn of the century, almost all deburring, edge contouring, and surface finishing work will be programmed into automated machining operations as a *primary material removal operation* and an integral part of the manufacturing process and systems. This will include large, medium, and small shops and manufacturing operations.

The change will be evolutionary, gaining momentum as new finishing tools are continually being developed, installed, and becoming a productive part of flexible manufacturing systems, computer automated machine tools, advanced machining systems, machine cells, special machines, and numerical and computer numerical controlled machine tools.

The future will be an excellent time for creativity, innovation, initiative, and opportunity for everyone associated with the manufacturing, machine tool, and material removal tool industries.

Index

Milton Keynes UK
Ingram Content Group UK Ltd.
UKHW031133141024
449569UK00006B/214